PHYLOGENETIC ANALYSIS OF MORPHOLOGICAL DATA

SMITHSONIAN SERIES IN COMPARATIVE EVOLUTIONARY BIOLOGY

The intent of this series is to publish innovative studies in the field of comparative evolutionary biology, especially by authors willing to introduce new ideas or to challenge or expand views now accepted. Within this context, and with some preference toward the organismic level, a diversity of viewpoints is sought.

Also in the series

Parascript
Parasites and the Language of Evolution
Daniel R. Brooks and Deborah A. McLennan

The Development and Evolution of Butterfly Wing Patterns
H. Frederick Nijhout

Hawaiian Biogeography
Evolution on a Hot Spot Archipelago
Edited by Warren L. Wagner and V. A. Funk

The Origin and Early Diversification of Land Plants
A Cladistic Study
Paul Kenrick and Peter R. Crane

Phylogenetic Analysis of Morphological Data

Edited by John J. Wiens

SMITHSONIAN INSTITUTION PRESS
Washington and London

Copy editor and production editor: Deborah L. Sanders
Designer: Janice Wheeler

Library of Congress Cataloging-in-Publication Data
Phylogenetic analysis of morphological data / edited by John J. Wiens.
p. cm. — (Smithsonian series in comparative evolutionary biology)
Includes bibliographical references.
ISBN 1-56098-841-x (alk. paper) — ISBN 1-56098-816-9 (pbk.)
1. Morphology. 2. Phylogeny. I. Wiens, John J. II. Title. III. Series.

QH351 .P58 2000
571.3—dc21 00-023910

British Library Cataloguing-in-Publication Data available

Manufactured in the United States of America
07 06 05 04 03 02 01 00 5 4 3 2 1

⊗ The paper used in this publication meets the minimum requirements of the American National Standard for Information Sciences—Permanence of Paper for Printed Library Materials ANSI Z39.48-1984.

CONTENTS

CONTRIBUTORS

Kevin de Queiroz
Department of Vertebrate Zoology
National Museum of Natural History
Smithsonian Institution
Washington, D.C. 20560

William L. Fink
Department of Biology and
Museum of Zoology
University of Michigan
Ann Arbor, Michigan 48109

David M. Hillis
Section of Integrative Biology
University of Texas
Austin, Texas 78712-1064

John P. Huelsenbeck
Department of Biology
University of Rochester
Rochester, New York 14627-0211

Paula M. Mabee
Department of Biology
University of South Dakota
Vermillion, South Dakota 57069

Lucinda A. McDade
Department of Ecology and
Evolutionary Biology and
Department of Plant Sciences
University of Arizona
Tucson, Arizona 85721

Steven Poe
Section of Integrative Biology and
Texas Memorial Museum
University of Texas
Austin, Texas 78712-1064

Bruce Rannala
Department of Ecology and Evolution
State University of New York
Stony Brook, New York 11794-6245

Donald L. Swiderski
Museum of Zoology
University of Michigan
Ann Arbor, Michigan 48109

John J. Wiens
Section of Amphibians and Reptiles
Carnegie Museum of Natural History
Pittsburgh, Pennsylvania 15213-4080

Miriam L. Zelditch
Museum of Paleontology
University of Michigan
Ann Arbor, Michigan 48109

PREFACE

One of the most important trends in the fields of evolution and ecology in recent years has been the growing appreciation for the importance of phylogenies. Biologists are increasingly realizing the necessity of a phylogenetic framework for understanding how organisms, their traits, and interactions between species evolve. The renewed interest in phylogenies has arisen partly from new molecular techniques and computer technologies, but perhaps most importantly, from the application of rigorous methods of phylogeny reconstruction to all types of data.

Most of our knowledge of the Tree of Life, both at lower and higher taxonomic levels, is based on phylogenetic studies of morphological data. Despite the growth of molecular systematics, morphology has been and will continue to be one of the most important and widely used sources of characters for phylogeny reconstruction. For example, morphology is the only way to address the phylogenies of fossil taxa and their relationships to living organisms. Similarly, many (if not most) extant species are known from only a handful of museum specimens and may never be studied using anything but morphology.

Use of morphological data in phylogenetic analysis presents a unique set of opportunities and problems, such as the inclusion of data from fossils, morphometrics, and ontogeny. Apart from these specialized types of data, there are also many fundamental issues in how morphological phylogenetics is practiced (e.g., character selection and delimitation) that are essentially unique to it. Yet, until now, there has not been a book-length treatment of specific topics in the phylogenetic analysis of morphology.

In many ways, morphological phylogenetics has been changing rapidly in recent years, with increasing competition and interaction with molecular systematics, the development of new methods, and the testing of long-standing assumptions. This volume provides an overview and summary of research on diverse topics relating to the phylogenetic analysis of morphological data. Admittedly, many topics are not fully covered (e.g., character independence, ordering), and many topics that are addressed could be the subjects of full-length books by themselves (e.g., hybrids, ontogeny). Furthermore, I expect that some of the chapters will be considered controversial, especially those based on a statistical or probabilistic framework. I hope that this book will highlight the extent to which some aspects of morphological phylogenetics have evolved in recent years (or need to evolve), the diversity of areas in which morphological phylogenetics has the potential to improve, and important issues for future research.

This book derives largely from a symposium entitled "Morphological Data in Phylogenetic Analysis: Recent Progress and Unresolved Problems," which was held in Saint Louis, Missouri, at the 1996 annual meeting of the Society of Systematic Biologists. I am grateful to the SSB for its financial support of the symposium. I thank Chuck Crumly for suggesting that I develop a book from the symposium and Peter Cannell for supporting this idea at Smithsonian Institution Press. Each chapter was refereed by two reviewers, and I am grateful to all who took the time to review one or more chapters: Phil Chu, Rob DeSalle, Vicki Funk, James Hanken, Richard Leschen, Bradley Livezey, Zhexi Luo, François Lutzoni, Paula Mabee, Charles Marshall, Emília Martins, James Rohlf, Gerald Smith, Peter Stevens, and John Wenzel.

DAVID M. HILLIS AND JOHN J. WIENS

1

MOLECULES VERSUS MORPHOLOGY IN SYSTEMATICS

Conflicts, Artifacts, and Misconceptions

In the 1960s, the newest area of biology—molecular biology—began making important contributions to one of the most established biological disciplines—systematics. Until then, the classification of life had been primarily based on morphological studies. Although there has been considerable cross-fertilization between morphological and molecular systematics, there has also been conflict of results and disagreement among practitioners of these two subfields. The popular and scientific presses have been quick to profile any new systematic study that purported to overturn some long-accepted taxonomic grouping, thus fueling the fires of disagreement. This has led to the perception of a battle of sorts between molecular and morphological systematists. But is the conflict real or spurious? How large is the common ground between the two schools? Are molecular studies completely reconfiguring the Tree of Life or merely making refinements and filling the gaps in our knowledge of it?

We see morphological and molecular approaches to systematics as having different strengths, but also as having many problems and difficulties of analysis in common. We believe that a coordinated effort between morphological and molecular systematists—or better yet, broader training of systematists so that they master both morphological and molecular analysis—is necessary to make real progress in assessing the world's biodiversity. Furthermore, we contend that significant incongruence between molecular and morphology-based phylogenies is rare: Most of the apparent conflict does not stand up to statistical scrutiny. Moreover, the two approaches to systematics are largely complementary, and each approach has advantages and disadvantages. Morphological and molecular studies can each address questions that cannot be addressed by the other.

And, where the two fields overlap, agreement of the results overwhelms any incongruity.

ADVANTAGES OF MOLECULAR DATA

The greatest advantage of molecular data in systematics appears to be the large number of observable characters available for analysis (Hillis 1987). Theoretical and empirical studies have shown that a sufficient number of characters is crucial in estimating phylogeny (e.g., figure 1.1; also, Hillis et al. 1994). Typical molecular studies involve several hundred or several thousand characters, and studies that include tens of thousands of characters even are becoming common (e.g., Bailey et al. 1992). This contrasts with morphological studies, in which a given analysis rarely includes more than a few hundred characters and averages about three characters per taxon (Sanderson and Donoghue 1989). The potential number of morphological characters may not be much greater than the number actually used in studies (at least not by more than a few orders of magnitude), but the potential number of molecular characters is much greater.

How might we estimate the number of useful molecular characters? Two basic assumptions of phylogenetic analysis limit the number of characters that can be used: (1) the characters are independent of one another, and (2) they are heritable. In other words, (1) change in one character being analyzed does not affect the probability of change in any other character being analyzed, and (2) the character variation passes (with rare mutation) from ancestors to descendants. The number of characters that fit these assumptions can be no larger than the number of nucleotide pairs in a genome (up to about 10^{12}) and in most cases will be much smaller (because of a large number of repeated sequences—and thus, a lack of independence). However, for most organisms, single-copy DNA sequences are abundant enough that the number of molecular characters available for analysis will rarely be limiting. For any given phylogenetic analysis, of course, only a subset of the available sequences are likely to be informative about the relationships of taxa under study.

Another advantage of molecular data is the wide range of substitution rates that exist across nucleotide sites. This permits phylogenetic analysis of everything from morphologically indistinguishable conspecifics to the most distantly related lineages of life—those with few or no morphological traits in common.

Molecular data also have the advantage that their genetic basis is usually known, whereas the genetic basis of most morphological characters is merely assumed. To date, few (if any) morphological phylogenetic findings have been discredited for being based on non-heritable variation. However, non-independence

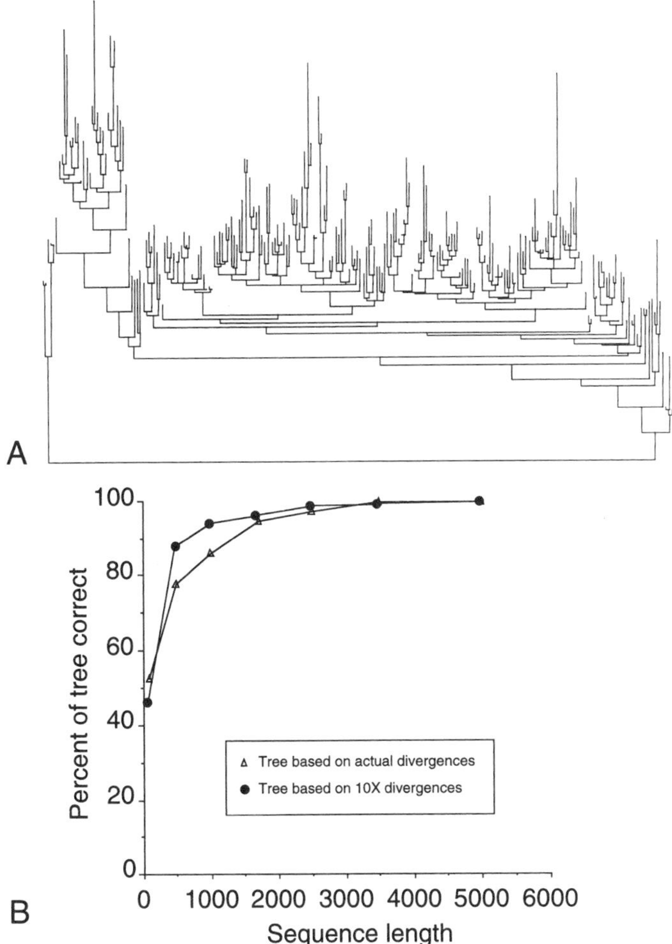

Figure 1.1. Increasing the accuracy of phylogenetic estimates—even in complex cases—by increasing the number of characters in the analysis. Adapted from Hillis (1998). **(A)** The model tree Hillis used in simulations was from an empirical study of angiosperm relationships by Soltis et al. (1997) (Kimura two-parameter model; gamma distribution of rate heterogeneity, with shape parameter set at 0.5). **(B)** Hillis based one simulation on the tree with the branch lengths that Soltis et al. estimated; another simulation was based on a tree of the same shape but with the branch lengths multiplied by 10. In either case, the accuracy of the inferred tree in a parsimony analysis increases with the number of examined characters up to a point, and all internal branches in the tree are correctly estimated with 5,000 nucleotides. "Percent of tree correct" was based on the partition metric (Robinson and Foulds 1981; Penny and Hendy 1985).

is likely a major problem in morphology, even though there are few good examples of non-independence in the literature (Emerson and Hastings 1998). A lack of knowledge regarding the genetic basis of morphological characters makes the potential for non-independence in morphology-based phylogenies a largely intractable problem. Although molecular data sets could likewise have the flaw of non-independence, discovering the flaw and finding appropriate alternatives (e.g., Dixon and Hillis 1993) are likely to be much easier than for morphological data sets.

Another advantage of the molecular approach is that characters can be selected and defined in a relatively objective manner. For example, a molecular systematist picks a gene and analyzes and reports all of its nucleotide sequences. Although alignment decisions and choice of the gene to be sequenced involve subjectivity, the criteria used to delimit characters typically are straightforward and objective. In morphological systematics, the characters must be discovered and delimited by the systematist, usually without any explicit criteria for character selection or coding. Thus, morphological data sets have the potential to be quite arbitrary (e.g., Gift and Stevens 1997). For example, morphologists do not generally report their criteria for including or excluding characters, and when criteria are given, they vary considerably among studies. (See chapter 2 of this book.) There is little to prevent morphologists from excluding those characters that cause their preferred hypothesis to be rejected, and morphological studies typically include too few characters anyway. These concerns would be greatly alleviated if morphologists routinely provided well-defined criteria for selecting and delimiting characters (as argued in chapter 2).

ADVANTAGES OF MORPHOLOGICAL DATA

The greatest advantage of morphological studies is that they allow for much more thorough taxonomic sampling than is possible with molecular analyses. Sampling a large number of taxa for molecular studies can be difficult because of the cost of sequencing, the need for relatively fresh material, the rarity (or extinction) of species, and the inaccessibility of the areas where certain taxa occur. In contrast, museum specimens can be used to cheaply and easily score many morphological characters for a vast array of taxa. Thorough taxon sampling is important for systematic revisions, studies of character evolution, and phylogenetic estimation. The greatest benefit of including additional taxa may be the potential to subdivide long branches in the estimated tree, which if insufficiently subdivided can result in inaccuracies in character reconstructions and in phylogenetic estimation (figure 1.2; also Hillis 1996, 1998). Long

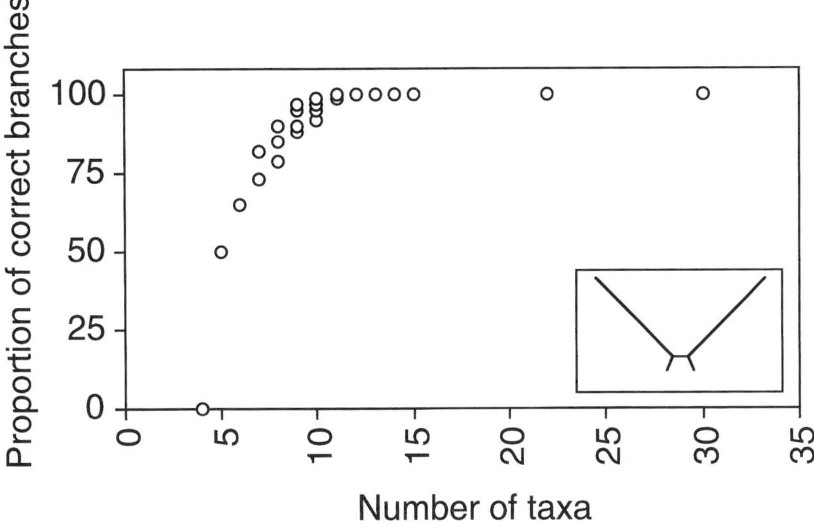

Figure 1.2. The importance of broad taxon sampling for a difficult phylogenetic problem. Adapted from Graybeal (1998). In Graybeal's simulations, the overall number of nucleotide positions analyzed was held constant as additional taxa were added to the parsimony analysis. In the simulation depicted here, each analysis covered a total of 40,000 nucleotides for a different number of taxa, from 4 to 30 (e.g., when 4 taxa were analyzed, each taxon was represented by 10,000 nucleotides; when 8 taxa were analyzed, each was represented by 5,000 nucleotides). The inset shows the starting tree for this analysis; the long branches have a length (p distance) of 0.5, and the short branches have a length of 0.05. Additional taxa were added so as to break up long branches evenly. Adding taxa—even at the expense of analyzing fewer characters per taxon—greatly improves the accuracy of the phylogenetic estimates.

branches can mislead all phylogenetic methods and thus can lead to answers that are strongly supported but wrong (Felsenstein 1978; Huelsenbeck 1995).

Morphology is also the only way that most fossil taxa can be analyzed phylogenetically. Extinct species not only represent a large proportion of Earth's biodiversity but also can be crucial for understanding the relationships among living taxa (Smith 1998). Fossils are important for correctly estimating relationships among living taxa in that they hold clues to many ancestral character states, and this information is needed to subdivide long branches among distantly related extant groups (Gauthier et al. 1988; Huelsenbeck 1991). The temporal information that fossils yield (i.e., estimated dates of appearance in the fossil record) can also be important for testing and rooting phylogenetic trees (e.g., Gauthier et al. 1988; Huelsenbeck 1994; Benton 1998; see also chapter 7 of this book).

Another advantage of morphology-based phylogenetics is that each morphological character is probably encoded by a different gene or set of genes, whereas in molecular data sets, many or all nucleotide characters may be drawn from the same gene. If the evolution of a gene differs from that of a species, trees reconstructed from molecular data may give well-supported wrong answers to questions about species phylogeny (Doyle 1992). Therefore, estimates of species phylogeny based on only one gene should be taken with caution. Morphological data sets are virtually immune to this problem and can be used as an important "reality check" to phylogenetic estimates that are based on the molecular characters of a single gene (Doyle 1992). Fortunately, most organisms' genomes contain thousands of unlinked genes, and the factors that cause gene and species lineages to differ are unlikely to systematically affect a large number of genes or a large number of taxa.

Finally, in addition to phylogeny reconstruction, morphology continues to play a crucial role in alpha taxonomy. In almost all groups of organisms, species are described and identified on the basis of morphological data. Therefore, even molecular systematists depend largely on morphology to determine which species they have sampled.

INCONGRUENCE AND CONFLICT

Apparently conflicting molecular and morphology-based trees are not hard to find in the phylogenetic literature. Yet on closer inspection, most of the incongruities prove to be spurious. We believe that the most common cause of incongruence between molecular and morphological phylogenies is simply weak support for either or both of the estimates, due to undersampling of characters and/or taxa. For example, when the molecular data set and the morphological data set for 40 species of phrynosomatid lizards were analyzed separately, the two resulting phylogenetic estimates differed in 74% of their clades; however, all of the conflicting nodes have low bootstrap or decay index values in one or both of the trees (Reeder and Wiens 1996). It is extremely easy to overestimate conflict if no attempt is made to evaluate support for clades. Examples of morphological and molecular phylogenies that conflict with each other but are both well supported are rare.

Applying different phylogenetic methods to the data (e.g., different optimality criteria or different search strategies for trees) may also lead to differences in phylogeny estimates (Hillis 1987). For example, two estimates of the relationships among sand lizard genera, one based on molecular data (Adest 1978) and the other based on morphological data (Etheridge and de Queiroz

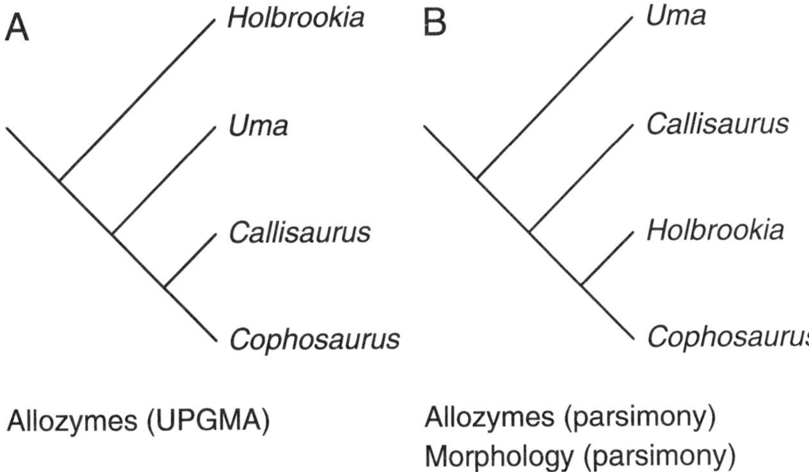

Figure 1.3. Apparent conflict between molecular and morphological data as an artifact of applying a different phylogenetic analysis method to each data set. **(A)** The phylogeny of the "sand lizard" clade of the family Phrynosomatidae based on allozyme data and UPGMA analysis (Adest 1978). **(B)** Sand lizard phylogeny based on morphology and parsimony analysis (Etheridge and de Queiroz 1988). The allozyme data used to construct tree A, when reanalyzed with the parsimony method, resulted in a tree identical to tree B (K. de Queiroz 1992).

1988), are completely different (figure 1.3), and the different types of data could be the reason. However, the molecular data had been analyzed with UPGMA (a method known to be highly sensitive to unequal branch lengths), whereas the morphological data had been analyzed using the parsimony criterion. Application of parsimony to the molecular data resulted in a tree that was identical to that estimated in the morphology study (K. de Queiroz 1992). In this case, the data sets cannot be said to be in conflict; it is the methods of analysis that disagree. Omland (1994) detected a similar reason for incongruence between estimates of duck phylogeny.

Another source of spurious conflict between findings of morphologists and molecular systematists is rooting. In morphology-based studies, rooting of the phylogenetic trees is often based on a broad sampling of living and fossil taxa, focused on the closest relatives (e.g., Wiley 1981; Maddison et al. 1984). In contrast, many molecular systematists use a single species (which may or may not be closely related to the group of interest) to root their trees, and this approach makes their estimates susceptible to long-branch problems. An excellent example of this is seen in estimates of whale phylogeny (figure 1.4). The molecular

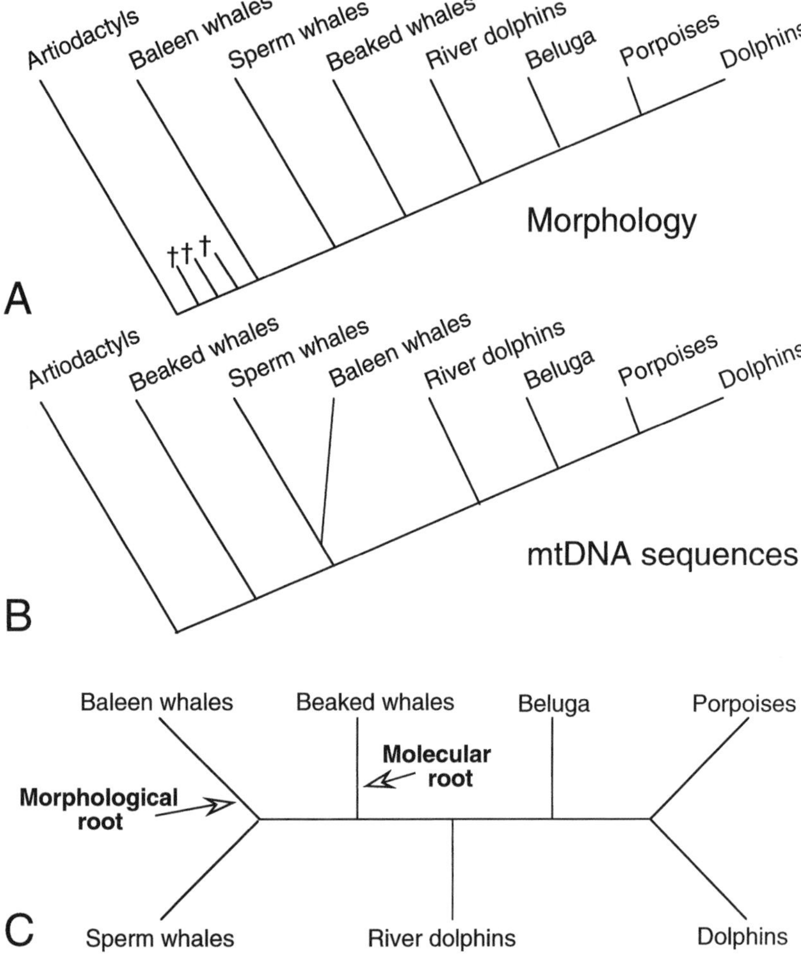

Figure 1.4. Apparent conflict between molecular and morphological data that is attributable to uncertain rooting of one of the phylogenetic trees (by use of a distant outgroup taxon). **(A)** The estimated phylogeny of the major whale taxa based on morphological data. The outgroups were three late Eocene fossil taxa (marked with daggers) that are closely related to modern whales but have some ancestral features such as pelvic limbs. Adapted from Messenger and McGuire (1998). **(B)** The estimated phylogeny of whales based on mitochondrial DNA sequences. Adapted from Milinkovitch et al. (1993, 1994). **(C)** Tree A and tree B superimposed. The two trees differ only in the position of the root. (Here, the "morphological root" is the root of tree A, and the "molecular root" is that of tree B.) Messenger and McGuire (1998) found that the position of the "molecular root" changed when the DNA data were reanalyzed with different combinations of extant artiodactyls used as the outgroup.

and morphological data sets yield different estimates of phylogeny: morphology supports the baleen whales as a natural group, whereas this group is paraphyletic in some molecular analyses (Milinkovitch et al. 1993, 1994). The molecular analyses, in turn, have led to major reassessments of morphological evolution (Milinkovitch 1995). Closer inspection shows that the tree based on morphology and the tree based on molecular studies have the same unrooted topology and differ only in the position of the root. Moreover, the root position suggested by morphology cannot be rejected by the molecular data, whereas the various root positions suggested by the molecular studies (in which rooting is dependent on outgroup selection) can be rejected by the morphological data (figure 1.4). In this case, the major difference between the two data sets appears to be the availability of closely related fossil outgroup taxa for morphological assessment (although there are also problems associated with the alignment of the molecular sequences; Cerchio and Tucker 1998; Messenger and McGuire 1998). Nonetheless, the problem is not limited to the inclusion or exclusion of fossil taxa; differences in taxon sampling among extant taxa may also lead to differences between molecular and morphology-based trees (Halanych 1998; Hillis 1998).

Conflict between phylogenies from molecular and morphological data sets may also be attributable to actual differences in phylogenetic histories—that is, when the phylogeny of the gene differs from the phylogeny of the species. Potential causes of mismatch between gene and species trees include paralogy, lineage sorting, and lateral transfer of genes between unrelated species (Doyle 1992; A. de Queiroz 1993). However, this problem appears to us to be uncommon (see also Brower et al. 1996).

A few cases of conflict between molecular and morphology-based phylogenies have defied explanation. In a recent review, Baker et al. (1998) identified seven studies that showed major differences between phylogenies based on morphological data sets and those based on molecular data sets. Recent estimates of crocodilian phylogeny are a well-documented example of such conflict (figure 1.5). Morphological and stratigraphic data strongly support placement of gavials *(Gavialis)* at the base of the crocodilian tree; these two types of data also support the idea that the morphological similarities in snout shape between gavials and false gavials *(Tomistoma)* are superficial and convergent. However, several different molecular data sets support the close relationship between these two genera.

Various lines of evidence seem to rule out differences in phylogenetic methods, rooting problems, long-branch attraction due to limited taxon sampling, different phylogenetic histories, and morphological convergence as possible explanations for this conflict. Given the discrepancy between molecular and mor-

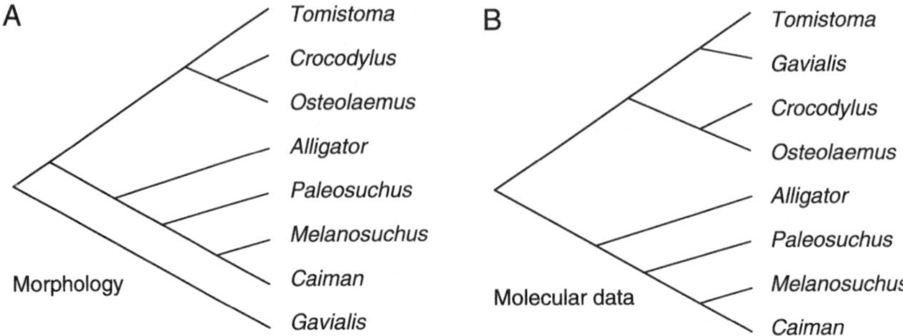

Figure 1.5. Conflict between morphological and molecular analyses of crocodilian relationships. Note the different placement of *Gavialis* in the two trees. Adapted from Poe (1996) and Brochu (1997). **(A)** Relationships estimated from morphological data. **(B)** Relationships estimated from various sources of molecular data.

phology-based phylogenetic estimates, Hass et al. (1992) called for a careful re-examination of the extant and fossil morphological data for crocodilians. Brochu (1997) did such an analysis and confirmed that the estimates based on the temporal record and morphological data are not consistent with the relationships suggested by the molecular data. Brochu (1997) echoed the sentiment of Hass et al. (1992)—that the data should be re-examined—but added that such a re-examination should include the molecular data as well. In this case, it is difficult to determine which data set is misleading and what the true phylogeny of the group may be.

MISCONCEPTIONS IN THE MOLECULES-VERSUS-MORPHOLOGY DEBATE

Perhaps the most common argument (e.g., Givnish and Sytsma 1997) for molecular and against morphological systematic studies concerns the issues of homology and homoplasy (convergence, parallelism, and reversals of character states). For instance, Hedges and Maxson (1996) have argued that "[m]orphological data are more susceptible to adaptive convergence . . . and therefore are less desirable for phylogeny estimation." They cited Doolittle (1994) in support of their position that molecular data are not susceptible to adaptive convergence. However, Doolittle (1994) merely argued that it is unlikely that similar proteins would have evolved from completely dissimilar ancestral sequences that underwent massive convergence, not that convergence per se is any less likely on a molecular level than on a morphological level. Doolittle (1994) did note some cases of molecular convergence but argued that in none of the cases was the ancestry of the proteins confounded as a result of the convergence. However, Doolittle (1994) also noted that the same was generally true of morphological convergence:

> We are all familiar with convergence at the morphological level, whereby unrelated or distantly related creatures come to look like each other, usually to the advantage of one or both. Ordinarily the resemblance is superficial and can be detected as such upon careful examination. Some marsupials look very much like certain carnivores, and porpoises and whales have found advantage in being shaped like fish, but the comparative anatomist has no trouble sorting out the marsupials from the placental mammals or the marine mammals from the fish.

In other words, convergent evolution of a large number of characters in distantly related taxa is no more a problem in molecular studies than it is in morphology, a point with which we agree.

The potential problem with homoplastic characters, on a molecular as well as a morphological level, becomes clear in an analysis of closely related taxa. In such cases, the convergence of characters need not be on a grand scale to overwhelm historical signal in the data. Similar selective conditions (say, high temperature and low rainfall) might well lead us to the incorrect conclusion that two species of desert oaks were more closely related to each other than to their actual respective sister species on nearby cool and wet mountain tops. But we would be just as likely to make this kind of error whether we examined leaf morphology or the sequence of an enzyme that was under strong selection for function at high temperatures. In either case, we could likely discover our error by examining other morphological attributes or other genes that were not under selection in response to temperature. In other words, not only the problem but also the solution is virtually the same for both molecular and morphological data.

That selection can be a confounding force for genotype as well as for phenotype can be easily demonstrated under experimental conditions. Cunningham et al. (1997) and Bull et al. (1997), using different viral systems, have provided two examples. Cunningham et al. (1997) showed that even complex deletion events can occur independently in response to selection for rapid growth, and that particular substitution events (in this case, producing stop codons) are likely to occur as a correlated response. One might easily conclude that such complex deletion events and correlated substitutions are strong evidence of common ancestry, but in this case they were actually responses to common selective and mutational conditions. In the study by Bull et al. (1997), several lineages of bacteriophage descending from the same ancestor were grown with different species of host bacteria. In a phylogenetic analysis of the entire genomes of the later generations, in all cases the lineages associated with the same host grouped together, even though they were no more closely related to each other than to any of the other lineages (figure 1.6). If a lineage grown on one host was switched to a second, then it often converged in sequence (and on the estimated phylogeny) to the lineages that had been grown on the second host all along. In fact, the true trees in these experiments were significantly worse as descriptions of the data than was the optimal tree (figure 1.6). Similar selective conditions led to similar sequences, which in turn led to misleading phylogenetic results. These studies demonstrate that selective convergence indeed can occur in molecular sequence data to the extent that it confounds phylogenetic analyses. Whether or not such convergence ever leads to phylogenetic errors in natural populations is just as much an open question for molecular data as it is for morphological data. However, we suspect that this sort of error is of little consequence (relative to sampling error, for instance), regardless of data type.

As a corollary to the argument regarding convergence, it is sometimes argued that molecular approaches avoid problems of homology determination (e.g., Sibley and Ahlquist 1987). However, molecular systematists must contend with problems of gene duplication, horizontal gene transfer, exon shuffling, questions of positional homology that arise from insertion and deletion events, and processes that result in concerted evolution of repeated gene sequences, in addition to the usual problems of convergence at the sequence level (Hillis 1994). Thus, although the problems are not identical for molecular and morphological data, homology assessment can be difficult for either kind of data in certain circumstances.

A second spurious argument for preferring molecular data over morphological data is that it is "circular" to use morphological data to construct phylogenies if we are interested in using the phylogenies to study morphological character evolution (e.g., Meyer et al. 1994; Hedges and Maxson 1996). This is a weak argument for ignoring morphological data, because most studies of character evolution involve only a few characters of interest, not the entire morphological data set (K. de Queiroz 1996; also chapter 8 of this book). The problem of circularity (or, more accurately, bias or influence) may apply to only a few characters of interest and will depend on the specific evolutionary question being addressed (K. de Queiroz 1996; also chapter 8). Furthermore, this argument applies to molecular data as well as morphology (Lee 1997).

A third misconception is the idea that molecular data will always provide an easy solution to problems that have proven difficult to resolve with morphology. Although this is sometimes true, groups that are poorly resolved by morphology are commonly difficult for molecular analyses as well, and branches that are well supported in molecular analyses are often the same ones that are

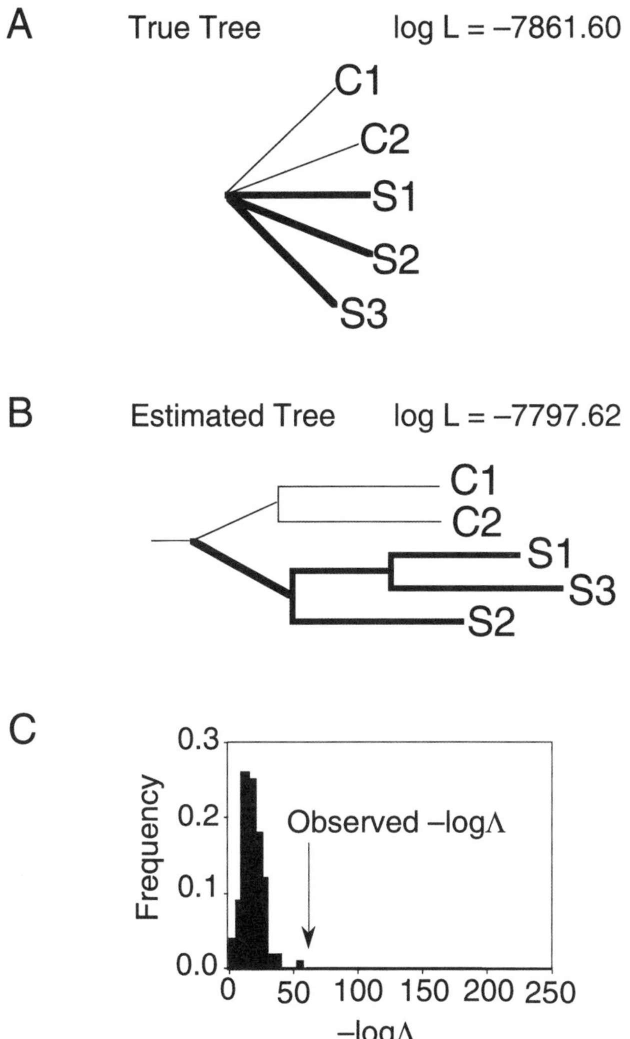

Figure 1.6. Similar selective conditions resulting in artificial grouping of molecular sequences. Adapted from Bull et al. (1997). **(A)** The true phylogeny of five viral lineages derived from bacteriophage ϕX174 under controlled conditions. All five lineages were derived directly from the wild-type ancestor but were grown in association with different host bacteria (C, *Escherichia coli* C; S, *Salmonella typhimurium*). **(B)** The best estimate of the phylogeny based on a maximum likelihood analysis of the complete DNA sequences of the five lineages. **(C)** A likelihood ratio test comparing tree A and tree B shows that the true tree (A) has a significantly poorer fit to the data.

well supported by morphology (e.g., Reeder and Wiens 1996). The most likely explanation for this pattern is that these difficult groups are ones that have speciated rapidly, such that there has been little time for either morphological or molecular changes to accumulate along branches (e.g., Kraus and Miyamoto 1991; Shaffer et al. 1997; Wiens and Reeder 1997). In the long run, we suspect that the resolution of such groups will come from sampling a very large number of molecular and morphological characters for a very large number of taxa, but the fundamental problem in these cases seems to be the rate of speciation and not the type of data. It is true that different data sets can be largely complementary by providing resolution at different parts of the tree (e.g., Hillis 1987; Pennington 1996), but this phenomenon is not restricted to a particular type of data. Two different molecular data sets, or two different morphological data sets, may also provide complementary resolution depending on the rate of evolution of the characters.

INTEGRATING MOLECULAR AND MORPHOLOGICAL DATA

Given that molecular data and morphological data have distinct advantages, can they be integrated to find the best estimate of phylogeny? One way to integrate these two types of data is through combined analysis—analyzing all the data simultaneously in a single matrix. The results of the combined analysis may then be compared with the results of each of the individual analyses—a process that is much like a standard statistical meta-analysis (Hillis 1995). In many circumstances, the combined analysis should yield a better estimate of phylogeny than any of the separately analyzed data sets yield, especially in cases where conflicts between trees from molecular and morphological studies are merely due to arbitrary resolutions of weakly supported clades. In such cases (which we believe to be the majority), combined analysis should improve the estimate by increasing the number of characters applied to the problem, and it may also reveal groups not seen in the trees from the separate data sets (Barrett et al. 1991; Chippindale and Wiens 1994).

In other circumstances, when molecular and morphological data sets strongly support different trees, combined analysis may or may not improve the estimate; the misleading data set may or may not overwhelm the other data set in the combined analysis (Bull et al. 1993; A. de Queiroz 1993). Several statistical approaches have been devised to test for significant incongruence among data sets (e.g., Rodrigo et al. 1993; Farris et al. 1994; Larson 1994; Lutzoni and Vilgalys 1995; Huelsenbeck and Bull 1996). When significant conflict is discovered, it is particularly important to investigate the causes of disagreement (Kluge 1989; Bull et al. 1993). Several potential causes have been outlined in this chapter. It should be noted that strongly supported conflicts may be limited to only a few taxa within the data sets and that the combined analysis may still provide an improved estimate for other parts of the tree in these cases (Wiens and Reeder 1997; Wiens 1998). Even if one chooses not to combine conflicting data sets, one should not simply ignore a particular hypothesis (molecular or morphology-based) on the grounds that one kind of data is generally superior to the other. Both should be considered viable hypotheses until the misleading data set can be identified.

CONCLUSIONS

Molecular and morphological approaches to systematics each have clear advantages; neither field can claim exclusive rights to the most informative data under all circumstances. If the goal of phylogenetics is to obtain the best estimate of the Tree of Life, reaching this goal will require cooperation and collaboration among and between morphological and molecular systematists. Furthermore, we believe that to train and label systematists as either morphological or molecular is to produce too many over-specialized graduates with a limited appreciation for the breadth and diversity of the field. All systematists should be just as comfortable examining and comparing museum specimens as they are at examining and comparing DNA sequences; likewise, they should be as comfortable collecting tissues for molecular analysis as preserving specimens for morphological study. There may always be some room for specialists, but we expect that the future will favor broader training and researchers who are not afraid to choose the best approach for any given phylogenetic puzzle—whether it means studying morphological or molecular variation, or both.

ACKNOWLEDGMENTS

We thank David Cannatella, Brad Livezey, François Lutzoni, Rob DeSalle, and Maria Servedio for comments on the manuscript.

REFERENCES

Adest, G. 1978. The relations of the sand lizards *Uma, Callisaurus,* and *Holbrookia:* An electrophoretic study. Doctoral dissertation, University of California, Los Angeles.

Bailey, W. J., K. Hayasaka, C. G. Skinner, S. Kehoe, L. C. Sieu, J. L. Slightom, and M. Goodman. 1992. Reexamination of the African hominoid trichotomy with additional sequences from the primate β-globin gene cluster. Molecular Phylogenetics and Evolution 1: 97–135.

Baker, R. H., Y. Xiaobo, and R. DeSalle. 1998. Assessing the relative contribution of molecular and morphological characters in simultaneous analysis trees. Molecular Phylogenetics and Evolution 9: 427–436.

Barrett, M., M. J. Donoghue, and E. Sober. 1991. Against consensus. Systematic Zoology 40: 486–493.

Benton, M. J. 1998. Molecular and morphological phylogenies of mammals: Congruence with stratigraphic data. Molecular Phylogenetics and Evolution 9: 398–407.

Brochu, C. A. 1997. Morphology, fossils, divergence timing, and the phylogenetic relationships of *Gavialis*. Systematic Biology 46: 479–522.

Brower, A. V. Z., R. DeSalle, and A. Vogler. 1996. Gene trees, species trees, and systematics: A cladistic perspective. Annual Review of Ecology and Systematics 27: 423–450.

Bull, J. J., J. P. Huelsenbeck, C. W. Cunningham, D. L. Swofford, and P. J. Waddell. 1993. Partitioning and combining data in phylogenetic analysis. Systematic Biology 42: 384–397.

Bull, J. J., M. R. Badgett, H. A. Wichman, J. P. Huelsenbeck, D. M. Hillis, A. Gulati, C. Ho, and I. J. Molineux. 1997. Exceptional convergent evolution in a virus. Genetics 147: 1497–1507.

Cerchio, S., and P. Tucker. 1998. Influence of alignment on the mtDNA phylogeny of Cetacea: Questionable support for a Mysiceti/Physeteroidea clade. Systematic Biology 47: 336–344.

Chippindale, P. T., and J. J. Wiens. 1994. Weighting, partitioning, and combining characters in phylogenetic analysis. Systematic Biology 43: 278–287.

Cunningham, C. W., K. Jeng, J. Husti, M. Badgett, I. J. Molineux, D. M. Hillis, and J. J. Bull. 1997. Parallel molecular evolution of deletions and nonsense mutations in bacteriophage T7. Molecular Biology and Evolution 14: 113–116.

de Queiroz, A. 1993. For consensus (sometimes). Systematic Biology 42: 368–372.

de Queiroz, K. 1992. Phylogenetic relationships and rates of allozyme evolution among the lineages of sceloporine sand lizards. Biological Journal of the Linnean Society 45: 333–362.

de Queiroz, K. 1996. Including the characters of interest during tree reconstruction and the problems of circularity and bias in studies of character evolution. American Naturalist 148: 700–708.

Dixon, M. T., and D. M. Hillis. 1993. Ribosomal RNA secondary structure: Compensatory mutations and implications for phylogenetic analysis. Molecular Biology and Evolution 10: 256–267.

Doolittle, R. F. 1994. Convergent evolution: The need to be explicit. Trends in Biochemical Sciences 19: 15–18.

Doyle, J. J. 1992. Gene trees and species trees: Molecular systematics as one-character taxonomy. Systematic Botany 17: 144–163.

Emerson, S. B., and P. A. Hastings. 1998. Morphological correlations in evolution: Consequences for phylogenetic analysis. Quarterly Review of Biology 73: 141–162.

Etheridge, R., and K. de Queiroz. 1988. A phylogeny of Iguanidae. Pages 283–368 *in* Phylogenetic Relationships of the Lizard Families (ed. R. Estes and G. Pregill). Stanford University Press, Stanford, Calif.

Farris, J. S., M. Källersjö, A. G. Kluge, and C. Bult. 1994. Testing significance of incongruence. Cladistics 10: 315–319.

Felsenstein, J. 1978. Cases in which parsimony or compatibility methods will be positively misleading. Systematic Zoology 27: 401–410.

Gauthier, J., A. G. Kluge, and T. Rowe. 1988. Amniote phylogeny and the importance of fossils. Cladistics 4: 105–209.

Gift, N., and P. F. Stevens. 1997. Vagaries in the delimitation of character states in quantitative variation: An experimental study. Systematic Biology 46: 112–125.

Givnish, T. J., and K. J. Sytsma. 1997. Homoplasy in molecular vs. morphological data: The likelihood of correct phylogenetic inference. Pages 55–101 *in* Molecular Evolution and Adaptive Radiation (ed. T. J. Givnish and K. J. Sytsma). Cambridge University Press, New York.

Graybeal, A. 1998. Is it better to add taxa or characters to a difficult phylogenetic problem? Systematic Biology 48: 9–17.

Halanych, K. M. 1998. Lagomorphs misplaced by more characters and fewer taxa. Systematic Biology 47: 138–146.

Hass, C. A., M. A. Hoffman, L. D. Densmore, and L. R. Maxson. 1992. Crocodilian evolution: Insights from immunological data. Molecular Phylogenetics and Evolution 1: 193–201.

Hedges, S. B., and L. R. Maxson. 1996. Molecules and morphology in amniote phylogeny. Molecular Phylogenetics and Evolution 6: 312–314.

Hillis, D. M. 1987. Molecular versus morphological approaches to systematics. Annual Review of Ecology and Systematics 18: 23–42.

Hillis, D. M. 1994. Homology in molecular biology. Pages 339–368 *in* Homology: The Hierarchical Basis of Comparative Biology (ed. B. K. Hall). Academic Press, San Diego.

Hillis, D. M. 1995. Approaches for assessing phylogenetic accuracy. Systematic Biology 44: 3–16.

Hillis, D. M. 1996. Inferring complex phylogenies. Nature 383: 130–131.

Hillis, D. M. 1998. Taxonomic sampling, phylogenetic accuracy, and investigator bias. Systematic Biology 48: 3–8.

Hillis, D. M., J. P. Huelsenbeck, and C. W. Cunningham. 1994. Application and accuracy of molecular phylogenies. Science 264: 671–677.

Huelsenbeck, J. P. 1991. When are fossils better than extant taxa in phylogenetic analysis? Systematic Zoology 40: 458–469.

Huelsenbeck, J. P. 1994. Comparing the stratigraphic record to estimates of phylogeny. Paleobiology 20: 470–483.

Huelsenbeck, J. P. 1995. The performance of phylogenetic methods in simulation. Systematic Biology 44: 17–48.

Huelsenbeck, J. P., and J. J. Bull. 1996. A likelihood ratio test to detect conflicting phylogenetic signal. Systematic Biology 45: 92–98.

Kluge, A. G. 1989. A concern for evidence and a phylogenetic hypothesis among *Epicrates* (Boidae, Serpentes). Systematic Zoology 38: 7–25.

Kraus, F., and M. M. Miyamoto. 1991. Rapid cladogenesis among the pecoran ruminants: Evidence from mitochondrial DNA sequences. Systematic Zoology 40: 117–130.

Larson, A. 1994. The comparison of morphological and molecular data in phylogenetic systematics. Pages 371–390 *in* Molecular Ecology and Evolution: Approaches and Applications (ed. B. Schierwater, B. Streit, G. P. Wagner, and R. DeSalle). Birkhauser Verlag, Basel, Switzerland.

Lee, M. S. Y. 1997. Molecules, morphology, and phylogeny: A response to Hedges and Maxson. Molecular Phylogenetics and Evolution 7: 394–395.

Lutzoni, F., and R. Vilgalys. 1995. Integration of morphological and molecular data sets in estimating fungal phylogenies. Canadian Journal of Botany 73: S649–S659.

Maddison, W. P., M. J. Donoghue, and D. R. Maddison. 1984. Outgroup analysis and parsimony. Systematic Zoology 33: 83–103.

Messenger, S. L., and J. A. McGuire. 1998. Morphology, molecules, and the phylogenetics of cetaceans. Systematic Biology 47: 90–124.

Meyer, A., J. Morrissey, and M. Schartl. 1994. Recurrent origin of a sexually selected trait in *Xiphophorus* fishes inferred from a molecular phylogeny. Nature 368: 539–542.

Milinkovitch, M. C. 1995. Molecular phylogeny of cetaceans prompts revision of morphological transformations. Trends in Ecology and Evolution 10: 328–334.

Milinkovitch, M. C., G. Orti, and A. Meyer. 1993. Revised phylogeny of whales suggested by mitochondrial ribosomal DNA sequences. Nature 361: 346–348.

Milinkovitch, M. C., A. Meyer, and J. R. Powell. 1994. Phylogeny of all major groups of cetaceans based on DNA sequences from three mitochondrial genes. Molecular Biology and Evolution 11: 939–948.

Omland, K. E. 1994. Character congruence between a molecular and a morphological phylogeny for dabbling ducks *(Anas)*. Systematic Biology 43: 369–386.

Pennington, R. T. 1996. Molecular and morphological data provide phylogenetic resolution at different hierarchical levels in *Andira*. Systematic Biology 45: 496–515.

Penny, D., and M. D. Hendy. 1985. The use of tree comparison metrics. Systematic Zoology 34: 75–82.

Poe, S. 1996. Data set incongruence and the phylogeny of crocodilians. Systematic Biology 45: 393–414.

Reeder, T. W., and J. J. Wiens. 1996. Evolution of the lizard family Phrynosomatidae as inferred from diverse types of data. Herpetological Monographs 10: 43–84.

Robinson, D. F., and L. R. Foulds. 1981. Comparison of phylogenetic trees. Mathematical Bioscience 53: 131–147.

Rodrigo, A. G., M. Kelly-Borges, P. R. Bergquist, and P. L. Bergquist. 1993. A randomization test of the null hypothesis that two cladograms are sample estimates of a parametric phylogenetic tree. New Zealand Journal of Botany 31: 257–268.

Sanderson, M. J., and M. J. Donoghue. 1989. Patterns of variation in levels of homoplasy. Evolution 44: 1671–1684.

Shaffer, H. B., P. Meylan, and M. L. McKnight. 1997. Tests of turtle phylogeny: Molecular, morphological, and paleontological perspectives. Systematic Biology 46: 235–268.

Sibley, C. G., and J. E. Ahlquist. 1987. Avian phylogeny reconstructed from comparisons of the genetic material, DNA. Pages 95–121 *in* Molecules and Morphology in Evolution: Conflict or Compromise? (ed. C. Patterson). Cambridge University Press, Cambridge.

Smith, A. B. 1998. What does palaeontology contribute to systematics in a molecular world? Molecular Phylogenetics and Evolution 9: 437–447.

Soltis, D. E., P. S. Soltis, D. L. Nickrent, L. A. Johnson, W. J. Hahn, S. B. Hoot, J. A. Sweere, R. K. Kuzoff, K. A. Kron, M. W. Chase, S. M. Swensen, E. A. Zimmer, S.-M. Chaw, L. J. Gillespie, W. J. Kress, and K. J. Sytsma. 1997. Angiosperm phylogeny inferred from 18S ribosomal DNA sequences. Annals of the Missouri Botanical Garden 84: 1–49.

Wiens, J. J. 1998. Combining data sets with different phylogenetic histories. Systematic Biology 47: 568–581.

Wiens, J. J., and T. W. Reeder. 1997. Phylogeny of the spiny lizards *(Sceloporus)* based on molecular and morphological evidence. Herpetological Monographs 11: 1–101.

Wiley, E. O. 1981. Phylogenetics: The Theory and Practice of Phylogenetic Systematics. John Wiley and Sons, New York.

STEVEN POE AND JOHN J. WIENS

2

CHARACTER SELECTION AND THE METHODOLOGY OF MORPHOLOGICAL PHYLOGENETICS

Estimation of phylogeny involves a series of interdependent steps: exploratory work (e.g., selection of characters and taxa), data collection and organization (e.g., character coding), and analysis (e.g., finding optimal trees). Many of the details and much of the rationale for choices made in these steps commonly are not made explicit in published papers; the analysis phase is often the only part of this series that is described (Thiele 1993). Morphological phylogenetics entails—in addition to those aspects that are characteristic of almost all phylogenetic analyses (such as using an algorithm to find a tree)—its own suite of usually implicit steps, such as the evaluation of character independence with reference to functional and growth constraints (Hennig 1966) and the methodology for coding observed anatomical variation for analysis (Pogue and Mickevich 1990).

Several researchers have noticed the lack of and stressed the need for explicitness in character selection criteria (e.g., Pimentel and Riggins 1987; Stevens 1991; Kesner 1994; Wiens 1995). However, the extent of this problem had not been formally studied and documented. Therefore, we did a systematic examination of more than a decade's worth of contributions to the primary literature of morphology-based phylogenetics. We investigated how morphologists chose characters for phylogenetic analysis, whether they were explicit about the criteria they set, and how much scientific merit the various character selection criteria have.

Table 2.1
Journals surveyed for morphological phylogenetic studies and the total number of applicable articles from each

Journal	No. articles
American Museum Novitates	71
Annals of the Entomological Society of America	25
Annals of the Missouri Botanical Garden	33
The Auk	11
Canadian Journal of Zoology	37
Cladistics	33
Condor	7
Copeia	22
Evolution	8
Herpetologica	8
Herpetological Monographs	4
Journal of Crustacean Biology	12
Journal of Herpetology	4
Journal of Mammalogy	3
Journal of Natural History	39
Journal of Zoology	11
Phycologia	4
Plant Systematics and Evolution	22
Systematic Botany	52
Systematic Zoology/Biology	13
Taxon	4
Wilson Bulletin	2
Zoological Journal of the Linnean Society	87

For each of the journals listed, our survey covered all issues dated 1986–1997, with the following exceptions: *American Museum Novitates*, 1995 volume not included; *Journal of Crustacean Biology*, 1996 and 1997 volumes not included.

METHODS

We collected data on morphological phylogenetic studies published in the years 1986–1997 in the 23 journals listed in table 2.1. To be included, an article had to be primary literature about a study of relationships, and the morphological characters had to have been selected for that study. We thus excluded papers that were reanalyses of data from the literature. In addition, we excluded studies in which trees were constructed by intuition; a numerical (but not necessarily computer-assisted) analysis of characters was required. Also excluded were studies in which the researchers did not seek independent characters (e.g., studies that involved principal-components analyses).

Table 2.2
Examples of criteria used in choosing characters

Reference	Criteria
Livezey 1986	"I chose to exclude mensural characters because . . . polarities and character states . . . are difficult to determine[.]"
Kraus 1988	"Two criteria were used for accepting characters for this analysis. First, the character had to vary discontinuously within the study group. Second, the character states had to be largely invariant within taxa. By 'largely invariant' I mean present in about 80% or more of the adult individuals of the species I examined. . . . Despite the second criterion, I occasionally allowed for the use of polymorphic characters in the analysis if the character was polymorphic for only one or two species and was invariant in the remaining species."
Cox and Urbatsch 1990	"Quantitative characters were generally rejected because the character states are circumscribed arbitrarily and because such arbitrary states probably have nothing to do with homology and analysis."
Bateman et al. 1992	Characters rejected when "known for less than ⅔ of OTUs."
Gensel 1992	Character excluded if "too variable (and either autapomorphic or homoplasious) to be included."
Hufford and Dickson 1992	"[C]haracters were deleted in the course of the investigation because of excessive missing data or polymorphism within groups."
Hood et al. 1994	"[C]haracters were selected for phylogenetic analysis because of their low intraspecific variation, presence in most taxa, ability to be scored unequivocally, and phylogenetic informativeness."
Kesner 1994	"[Desired] characters distinctive, biologically significant, and sufficiently constant within OTUs to allow for reliable character state assignment."

Each study that met our requirements for inclusion was evaluated as to whether or not criteria for character selection were mentioned, and any criteria were noted. Specifically, we scored papers positively if they mentioned exclusion of characters for any reason (other than to fit the two basic assumptions noted below), and we scored papers negatively if they did not mention character selection or if they explained character selection only in terms of which types were included. We assumed that independence and putative homology (common ancestry of states) are a part of any character selection strategy and therefore gave no credit for statements to this effect. Examples of character selection criteria are exclusion of a character because of high levels of intraspecific variation and exclusion because of missing data. Table 2.2 gives examples from the literature.

It is possible to distinguish between operational and conceptual criteria. For example, requiring "low" intraspecific variation is a conceptual criterion, whereas allowing only characters that have fixed states in terminal taxa is an operational criterion. However, because so few of the papers (<1%) included operational criteria, we chose to combine these two subcategories. It is also possible in some cases to distinguish between a criterion and a justification for that criterion. For example, a researcher could choose to reject all quantitatively coded characters (a criterion) because the continuous nature of such characters is thought to be unsuitable for cladistic analysis (a justification). However, justifications can be used as criteria, and vice versa, depending on context. So we treated criteria and justifications as one category in our scoring.

Although we focused on the methods section of articles in looking for the general character sampling schemes, we also examined the character descriptions. Some authors gave no general sampling scheme but did include detailed discussions of variation and of decisions concerning particular characters. Such discussions were scored positively for offering criteria. However, given the large number of articles we surveyed, we may have missed a few listings of criteria, particularly if they were buried in lengthy character descriptions.

We used the data that we collected to investigate four possibilities. First, we determined whether morphologists generally were explicit about character selection criteria. Second, we divided the data by journal type to assess whether authors publishing in general systematic journals were more conscientious than those publishing in taxon-oriented journals. Third, we compared our data on plant studies with our data on animal studies to detect any differences between the two fields. Fourth, we mapped trends over time to assess whether the advancement of phylogenetic methods has been paralleled by an increase in documented rigor of character selection.

RESULTS AND DISCUSSION

A Lack of Explicitness

The results of our survey suggest that, in general, morphologists are not explicit about how they choose systematic characters: Of the 512 morphological phylogenetic studies examined, 102 (20%) mentioned criteria for exclusion of char-

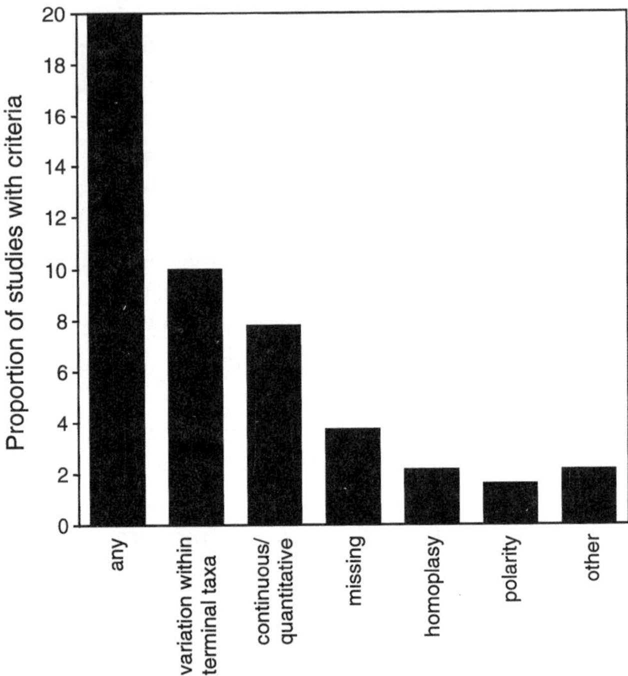

Figure 2.1. Explicitness of the morphological phylogenetic studies in our survey and the frequency with which the various character exclusion criteria were used. "Any" (meaning "any criterion for character exclusion was mentioned") is the overall percentage of studies that were explicit about character exclusion criteria. Some studies listed more than one criterion. "Other" refers to criteria that matched none of the others listed.

acters (figure 2.1). Papers in general systematics (or conceptual) journals were no more explicit than papers in taxon-oriented (organismal) journals, nor was there a difference between animal and plant studies ($P > 0.10$ for both comparisons, G-test of independence; figure 2.2). Furthermore, morphological phylogenetics has not shown a consistent trend toward more explicitness over the years ($P > 0.10$, Spearman's coefficient of rank correlation; figure 2.3).

Why might this state of affairs be maintained, despite pleas in the literature for change (e.g., Pogue and Mickevich 1990; Wiens 1995)? One possible explanation (which could not be addressed in a literature survey) is that morphological systematists believe that a common filter exists among them (Thiele 1993), such that discovering morphological characters is an objective and repeatable process even without listed criteria. Such commonality of observation seems unlikely, especially considering that different observers set dissimilar

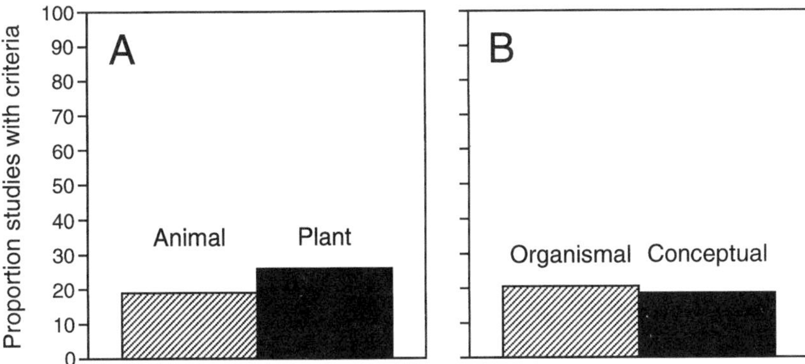

Figure 2.2. Explicitness of morphological phylogenetic studies according to type of journal in which they were published: **(A)** animal-studies or plant-studies journal; **(B)** taxon-oriented (organismal) or general systematic journal.

character state boundaries when presented with the same data on variation (Gift and Stevens 1997). Although there is probably little disagreement over what constitutes an ideal character, there is evidence of greatly differing opinions among morphologists as to where the dividing line is between a marginally acceptable and unacceptable character. The results of our survey suggest that people select characters differently. Comparing studies of the same organisms by different researchers is also telling (e.g., compare Good 1988 with J. A. Campbell and Frost 1993).

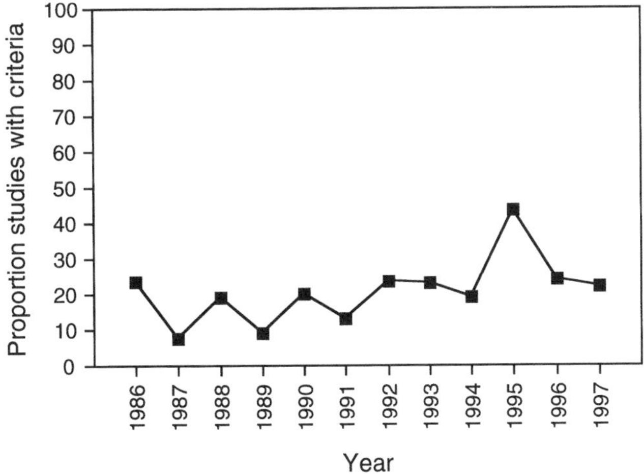

Figure 2.3. Explicitness of morphological phylogenetic studies by year of publication.

Inconsistency of character selection criteria can occur even within a single study. Stevens (1991) pointed out several examples of researchers who claimed to reject "quantitative" characters but who nevertheless accepted characters that could be described with an arbitrary quantitative cutoff, such as "calyx: >5 mm [0], <5 mm [1]" (or, less obviously, "calyx: large [0], small [1]"). A common filter is a myth.

Why Be Explicit?

One reason to be explicit about character selection criteria is simply to increase the rigor of morphological systematics. We recognize that an element of subjectivity may always be present in character selection—no two systematists are likely to view variation in exactly the same way. But minimizing subjectivity in methods is a goal of science. If the selection of characters remains a "black box" between selecting taxa and forming a matrix of zeros and ones, morphological phylogenetics will continue to be vulnerable to attack from those who accuse researchers of manipulating data to reach a preconceived phylogeny (e.g., Pritchard 1994) and from those who see such methods as sound in theory but flawed in implementation (e.g., Gould 1983). There is evidence that preconceived ideas about relationships affect the character choice of some systematists (e.g., those who screen out characters that show "too much homoplasy"; see below). The extent to which characters are picked to produce the desired tree is open to question if no selection criteria are stated. Furthermore, the differences between phylogenetic trees constructed by different authors for the same taxa may be difficult to interpret without knowing what criteria the authors used for including and excluding characters.

A second reason to be explicit is to allow for testing of the validity of character selection criteria and the properties of particular types of characters. For example, intraspecific variation is one of the most widely cited criteria for excluding characters, but the merit of this criterion has been tested only recently (e.g., J. A. Campbell and Frost 1993; Wiens 1995, 1998b; Wiens and Servedio 1997). Attempts to analyze the properties of polymorphic characters have been hampered by the fact that many morphological data matrixes do not indicate any intraspecific variation for the taxa and characters coded. Without explicit criteria for character selection, it is impossible to tell whether the absence of polymorphism in a matrix is due to actual rarity of intraspecific variation, the researcher's exclusion of polymorphic characters, or the researcher's deliberate failure to report uncommon variants within species (Wiens 1995). Even if polymorphic characters are explicitly included, many characters may still be excluded because "too much" intraspecific variation was observed (e.g.,

J. A. Campbell and Frost 1993). Thus, without explicit criteria for character selection, it is difficult to tell whether or not a given morphological study provides an unbiased sample of polymorphic characters, and what the nature and extent of the biases (if any) are. Because of this lack of information, empirical studies of polymorphic characters have so far been based on only a handful of data sets that included an explicit and seemingly unbiased sampling protocol for intraspecifically variable characters. A comparable situation exists for quantitative characters (Thiele 1993).

Similarly, comparisons of homoplasy (or some other measure) between different character "types" may be confounded by the methods by which an investigator chooses particular types of characters. For example, the finding of de Queiroz and Wimberger (1993) that behavioral characters are comparable in homoplasy levels to morphological characters could be explained by at least two hypotheses. First, there may be some inherent similarity in evolutionary plasticity in these two classes of data. Alternatively, systematists may be evaluating behavioral and morphological data through much the same filter, such that they tend to choose only characters that vary in an acceptable way, regardless of data type. The interpretation of comparative results such as these is crucial: the first of these hypotheses tells us something about evolution, whereas the second tells us something about systematists. These hypotheses can be tested only if investigators are explicit about character selection criteria.

Comparisons of rates of morphological evolution also may be rendered uninterpretable if no criteria for accepting or rejecting characters are listed. For example, Slowinski (1995) used seven characters in a study of 18 species of coral snakes and concluded that the group is "morphologically conservative." In contrast, Kluge (1989) used 53 morphological characters in his analysis of the relationships of 10 boid snakes. Does the discrepancy in number of characters included indicate that coral snakes are less variable than boas? It might, or it might mean only that Kluge and Slowinski have very different filters for character acceptability. We cannot test either of these explanations because neither of these authors gave explicit general criteria for how they chose characters.

Character Selection Criteria

In this section, we discuss the merits of various criteria for selecting morphological phylogenetic characters. Each of these criteria was mentioned by several authors of morphological phylogenetic papers; criteria listed only once are not discussed. In evaluating these criteria, our main concern is whether use of each criterion is likely to contribute to greater accuracy in morphological phylogenetics (where accuracy is the probability of estimating the true organismal phylogeny). However, we acknowledge that in most empirical studies the true phylogeny is unknown, that our predictions about accuracy come largely from statistical analyses, computer simulations, and congruence studies (see Hillis 1995 for a review), and that not everyone is concerned about phylogenetic accuracy (e.g., Kluge 1997; Siddall and Kluge 1997) or convinced by studies that attempt to address it.

In discussing these criteria, we point out some perceived disadvantages of particular character types that in fact appear to be applicable to morphological phylogenetics in general. Nonetheless, these potential problems should not be taken as indictments of the quality of morphological data relative to other types of data, for three main reasons. First, as discussed below, most of the perceived disadvantages are unfounded for morphology in general, as well as for particular types of morphological data. Second, some of the disadvantages discussed here are the same for or have clear analogs in other types of data. For example, coding intraspecific variation can pose problems regardless of the data type. Third, although some problems are for the most part limited to morphology (e.g., continuous variation), other data types have their own problems as well (e.g., alignment in DNA sequence data).

Some of the criteria discussed below overlap in content or in the sense that one criterion may be used as a justification for another. Our categories are thus somewhat arbitrary, and discussions of some criteria may bear on others.

Variation within terminal taxa. We found that the most common reason given for excluding characters was variation within terminal taxa (figure 2.1), particularly variation within species (i.e., polymorphism). Although there is evidence that polymorphic characters are more homoplastic than fixed characters (J. A. Campbell and Frost 1993; Wiens 1995), there is also evidence (from the same data sets) that polymorphic characters contain useful phylogenetic information. These two observations appear to make for a difficult choice in terms of whether or not to include intraspecifically variable characters (i.e., is it better to include more characters despite the increase in homoplasy or fewer characters with less homoplasy?). However, studies of real and simulated data sets (Wiens and Servedio 1997; Wiens 1998b) suggest that the choice is clear: Analyses that include polymorphic characters are consistently more accurate than those that exclude them, even when the data sets show a strong positive relationship between levels of homoplasy and intraspecific variability. Even though intraspecifically variable characters do seem to be more homoplastic, they evidently contain information that more than compensates for their homoplasy.

Similarly, characters are often excluded because they vary among the species within a higher level terminal taxon. In a simulation study, Wiens (1998a) showed that excluding such characters consistently decreased the accuracy of phylogenetic estimates, relative to including such characters. Furthermore, as discussed in more detail in chapter 5 of this book, Wiens's analysis of published morphological data for squamate reptiles (from Estes et al. 1988) suggested that (1) characters that vary among the species of a given higher taxon hold significant phylogenetic information for recovering higher level relationships (but are somewhat less informative than characters that are fixed within the higher taxon) and (2) there is generally no significant relationship between levels of homoplasy and the extent to which characters vary among the species within a given higher taxon. Thus, there is little support for the practice of excluding characters merely because they vary within the terminal units of a phylogenetic analysis.

Missing data. The desire to avoid having "missing data entries" (unknown or unscorable character states for some taxa) is sometimes used as a justification for excluding characters (figure 2.1). A recent simulation study (Wiens 1998c) provided little support for this practice. Under a wide variety of simulated conditions, adding characters with missing data significantly increased the phylogenetic accuracy of a given data set. Therefore, excluding characters with missing data must significantly decrease accuracy. Even when 75% of the data cells are scored as unknown, including such characters can still significantly increase accuracy.

Missing data entries themselves appear to be relatively harmless. However, taxa with missing data for a given character are effectively unsampled for that character, and limited taxon sampling may lead to positively misleading results through long branch attraction (Swofford et al. 1996). Under some conditions, including characters that cannot be scored for many taxa may decrease phylogenetic accuracy. This result seems most likely when very few taxa are sampled and the taxa that are sampled are distributed on the true phylogeny such that there are long terminal branches connected by a short internal branch (i.e., the situation described by Felsenstein 1978). Nevertheless, when unknown data entries are distributed randomly among taxa or are confined to a monophyletic subset of taxa, inclusion of characters with incomplete data is much more likely to increase accuracy than decrease it (Wiens 1998c). Therefore, we reject the general practice of excluding characters simply because they have missing data.

Continuous and quantitatively coded variation. Many authors who gave criteria mentioned a desire to avoid "continuous" variation in favor of "discrete" variation and to avoid "quantitative" coding in favor of "qualitative" coding. The terms "continuous" and "quantitative" often are used interchangeably but have distinct definitions (Thiele 1993). "Qualitative" and "quantitative" are ways of describing variation (descriptively and numerically, respectively) rather than properties of variation; "continuous" and "discrete" are potential properties of frequency distributions. Many types of variation can be coded either qualitatively or quantitatively, and many qualitatively coded data are "based on a quantitative phenomenological base filtered through the reified semantic discontinuities of . . . [anatomical] terminology" (Stevens 1991). Although these terms are distinct, we discuss them together because the reported justifications for exclusion of each of these character types are similar.

Quantitative coding and continuous variation have been rejected for any of a variety of reasons: because continuous variation is theoretically unsuitable for cladistic analysis (e.g., Pimentel and Riggins 1987), because state delimitation is arbitrary (e.g., Cox and Urbatsch 1990), because there is no consensus on a suitable coding method for quantitative data (see reviews listed below), because some types of continuous variation are difficult to characterize objectively (e.g., Wiens 1993), or because accuracy is reduced by use of such characters (e.g., C. S. Campbell 1986). These criticisms, which in general are poorly founded, are each discussed below.

Certainly phylogenetic methods for morphology were designed with discrete variation in mind (Hennig 1966; Wiley 1981). However, it does not follow from this intention that useful phylogenetic information cannot be extracted from continuous variation. In fact, numerous methods have been devised to code continuous variation in ways that satisfy cladistic assumptions of independence and discreteness of states (e.g., Archie 1985; Chappill 1989; Thiele 1993; Strait et al. 1996). Furthermore, it is not clear just how discrete characters must be to be considered useable. Stevens (1991) gave examples of variation that could be justifiably coded with either discrete cutoffs or more quantitative methods and showed that characters cannot be sharply distinguished as either discrete or continuous; rather, they fall along a continuum (see also figure 2 of Thiele 1993). There appears to be nothing uniquely undesirable about continuous variation in phylogenetic analysis, and there seems to be no objective means to distinguish "bad" ("continuous") variation or "good" ("discrete") variation in practice. Furthermore, Thiele (1993) has shown that quantitative characters can be comparable to more traditionally scored characters in homoplasy levels, phylogenetic informativeness, and resulting tree topology.

If arbitrary delimitation of states is a problem at all, it is for phylogenetic analysis of morphology in general, not just for analysis of quantitative data. Arbitrary delimitation of states is common in qualitatively coded variation, as in the decision to code a size difference as "large (0), small or absent (1)," or as "present (0), absent (1)," or as "large (0), small (1), absent (2)." In presence/

absence coding, which might be viewed as nonarbitrary, variation in size and shape is ignored under the rubric of "present." Is the presence/absence division really more significant, evolutionarily, than the large/small division? In fact, quantitatively coded variation may be less subject to vagaries in coding than is qualitatively coded variation, because explicit statistical or methodological grounds are often used in quantitative coding (Thiele 1993).

Although we do not advocate use of characters for which state delimitation is completely arbitrary, we recognize that stochastic genetic and environmental variation will produce situations where a "correct" coding scheme is not obvious but where potentially useful phylogenetic information remains. An example of so-called arbitrary coding that does not necessarily mislead is seen in Stevens (1991). His table 1 depicts a set of states coded in four different ways. He argued from this example that such characters are not suitable for cladistic analysis because there is no nonarbitrary way of deciding which coding scheme is best. However, although the different coding schemes do not produce identical groupings, the various possible relationships are largely compatible and in many cases differ only in the degree of resolution. For example, three of the four coding schemes group taxa A, B, and C together, three of four group A′, B′, and C′ together, and all find the maximum difference (e.g., one step) between taxa A and C′. Methods such as gap weighting (e.g., Thiele 1993) and frequency coding (e.g., Wiens 1995), which give greater weights to larger differences between taxa and lower weights to differences between similar taxa should be useful in cases where state delimitation has a potentially large arbitrary component.

The lack of a consensus regarding quantitative coding methods is well documented (see Archie 1985; Chappill 1989; Farris 1990; Strait et al. 1996). However, no accepted protocol exists for coding qualitative characters either, and examples of varying methods for coding discrete or qualitative variation abound (e.g., binary versus multistate, composite versus reductive; Wilkinson 1995). Clearly, the lack of a consensus for coding methods is not restricted to quantitative characters.

In our survey, a few authors rejected characters because of difficulty in objectively assigning character states (e.g., when there was continuous variation in shape or color among taxa). We argue that even if characters cannot be coded into discrete character states, they still can be characterized quantitatively in a phylogenetically useful way. For example, the thin-plate spline method may be useful in converting shape differences into phylogenetic characters (Zelditch et al. 1995) and one could conceivably assign numerical values to shades of color. However, some characters may be intractable to both quantitative and qualitative coding—for example, continuously varying shape characters lacking obvious landmarks for making either qualitative character state boundaries or measurements. For this reason, we consider difficulty in characterization to be a legitimate but perhaps overly used reason to reject characters.

The idea that inclusion of quantitative characters decreases phylogenetic accuracy has yet to be addressed directly. However, an empirical study by Thiele (1993) showed that continuous characters (at least when coded using gap weighting) can display significant nonrandom covariation and can result in trees that are significantly similar to or identical to trees based on qualitative characters. We take these results as evidence that including continuous characters increases accuracy, and we know of no examples that support the opposite conclusion.

Unknown polarity. At one point in the history of phylogenetics, it was thought that the plesiomorphic state of a character had to be determined a priori for that character to be useful in phylogenetic analysis (e.g., Wiley 1981). Consequently, many authors rejected characters that could not be polarized, and some have given this as an explicit criterion for excluding data. However, the idea that characters must be polarized to be included in a phylogenetic analysis has been shown to be misguided (e.g., Swofford and Olsen 1990). If an analysis includes a hypothetical ancestor taxon or outgroup to root the ingroup tree, then an unpolarized character merely represents a character with an ambiguous or unknown state for that taxon and still contributes to the resolution of ingroup relationships. Simulations suggest that excluding characters from analysis simply because their states cannot be determined in a few taxa significantly decreases phylogenetic accuracy (Wiens 1998c). Furthermore, unknown polarity of the characters is not likely to be an issue in a study design that includes outgroup taxa as terminal units in a global phylogenetic analysis (as advocated by Nixon and Carpenter 1993), in contrast to a design that requires a priori evaluation of character polarity (e.g., Maddison et al. 1984).

Level of homoplasy. Some authors stated that they chose characters that are conservative or not too homoplastic. This practice is flawed in that our perception of whether a character is homoplastic or not depends on some a priori notion of phylogeny (Hennig 1966; Wiley 1981), and the authors citing "too much homoplasy" as a criterion for character exclusion did not state how they determined these characters to be too homoplastic. Nevertheless, choosing characters that evolve at an appropriate rate for the level of analysis has become recognized as an important step in phylogenetic analyses of molecular data (Hillis et al. 1996). Practical means have been developed to identify and address these issues for molecular sequence data (e.g., comparing types of changes, such as transitions versus transversions; differential weighting to accommodate rate heterogeneity; identifying hypervariable regions of alignment), but many of these practices do not have clear analogs for morphological data. Researchers attempting to identify general classes of morphological characters that are too homoplastic for inclusion in phylogenetic studies have consistently arrived at the same conclusion: that these questionable character types do contain useful phylogenetic information and thus should not be excluded (e.g., integumentary characters in birds [Chu 1998]; reductive characters [Begle 1991; Buckup 1993]; morphometric characters [Thiele 1993]; and polymorphic characters [J. A. Campbell and Frost 1993; Wiens 1995, 1998b]). Until certain classes of morphological characters are convincingly identified as generally misleading, or until a method is developed to consistently identify misleading characters in particular cases, we reject the practice of excluding "excessively homoplastic" characters.

SUMMARY AND RECOMMENDATIONS

Our main conclusion from our survey is that most morphologists are not explicit about how they choose systematic characters. Our most basic recommendation from this work is not controversial: quite simply, morphological systematists need to discuss character selection and give their operational criteria for rejecting characters. The methodology of winnowing observed morphological variation into a data matrix of character states should be given as standard practice, and reviewers and advisers should ask questions about this methodology as a matter of course. The lack of explicit methodology for character selection for most morphological data sets seriously compromises (1) the claimed objectivity of modern morphological phylogenetics, (2) the comparison of phylogenies published by different authors, (3) tests of the properties of different types of morphological characters, (4) the comparison of morphological data to other types of characters (e.g., molecular, behavioral), and (5) the comparison of morphological evolution between taxonomic groups.

We found that systematists use a variety of criteria to exclude characters. However, there is little evidence to support the validity of most of the commonly cited criteria. In fact, simulation and congruence studies suggest that application of many of these criteria may actually decrease phylogenetic accuracy by discarding useful data. Although our discussion of character exclusion criteria has been largely from a perspective of maximizing phylogenetic accuracy, we suggest that a philosophically based "total evidence" approach to phylogenetics (e.g., Kluge 1989) would likewise require abandoning these exclusion criteria.

How should morphologists choose characters? The lack of evidence to justify most of the listed character rejection criteria suggests that much more variation could be included in phylogenetic analyses than is used presently. We hope that this review will encourage systematists to be more explicit about their methodology and to consider carefully their exclusion of potentially informative data.

ACKNOWLEDGMENTS

Thanks to Peter Stevens and Brad Livezey for helpful comments in review. S.P. was supported by fellowships from the National Science Foundation and the Smithsonian Institution.

REFERENCES

Archie, J. W. 1985. Methods for coding variable morphological features for numerical taxonomic analysis. Systematic Zoology 34: 326–345.

Bateman, R., W. A. di Michele, and D. Willard. 1992. Experimental cladistic analysis of anatomically preserved arborescent lycopsids from the Carboniferous of Euramerica: An essay on paleobotanical phylogenetics. Annals of the Missouri Botanical Gardens 79: 500–559.

Begle, D. P. 1991. Relationships of the osmeroid fishes and the use of reductive characters in phylogenetic analysis. Systematic Zoology 40: 33–53.

Buckup, P. A. 1993. Phylogenetic interrelationships and reductive evolution in neotropical characidiin fishes (Characiformes, Ostariophysi). Cladistics 9: 305–341.

Campbell, C. S. 1986. Phylogenetic reconstructions and two new varieties in the *Andropogon virginicus* complex (Poaceae: Andropogoneae). Systematic Botany 11: 280–291.

Campbell, J. A., and D. R. Frost. 1993. Anguid lizards of the genus *Abronia:* Revisionary notes, descriptions of four new species, phylogenetic analysis, and key. Bulletin of the American Museum of Natural History 216: 1–121.

Chappill, J. A. 1989. Quantitative characters in phylogenetic analysis. Cladistics 5: 217–234.

Chu, P. C. 1998. A phylogeny of the gulls (Aves: Larinae) inferred from osteological and integumentary characters. Cladistics 14: 1–43.

Cox, P. B., and L. Urbatsch. 1990. A phylogenetic analysis of the coneflower genera (Asteraceae: Heliantheae). Systematic Botany 15: 394–402.

de Queiroz, A., and P. Wimberger. 1993. The usefulness of behavior for phylogeny estimation: Levels of homoplasy in behavioral and morphological characters. Evolution 47: 46–60.

Estes, R., K. de Queiroz, and J. A. Gauthier. 1988. Phylogenetic relationships within Squamata. Pages 119–281 *in* Phylogenetic Relationships of the Lizard Families (ed. R. Estes and G. K. Pregill). Stanford University Press, Stanford, Calif.

Farris, J. S. 1990. Phenetics in camouflage. Cladistics 6: 91–100.

Felsenstein, J. 1978. Cases in which parsimony or compatibility methods will be positively misleading. Systematic Zoology 27: 401–410.

Gensel, P. 1992. Phylogenetic relationships of the zosterophylls and lycopsids: Evidence from morphology, paleoecology, and cladistic methods of inference. Annals of the Missouri Botanical Gardens 79: 450–473.

Gift, N., and P. F. Stevens. 1997. Vagaries in the delimitation of character states in quantitative variation: An experimental study. Systematic Biology 46: 112–125.

Good, D. A. 1988. Phylogenetic relationships among gerrhonotine lizards. University of California Publications in Zoology 121: 1–138.

Gould, S. J. 1983. Hen's Teeth and Horse's Toes: Further Reflections in Natural History. W. W. Norton, New York.

Hennig, W. 1966. Phylogenetic Systematics. University of Illinois Press, Urbana.

Hillis, D. M. 1995. Approaches for assessing phylogenetic accuracy. Systematic Biology 44: 3–16.

Hillis, D. M., B. K. Mable, and C. Moritz. 1996. Applications of molecular systematics: The state of the field and a look to the future. Pages 515–543 *in* Molecular Systematics, 2nd edition (ed. D. M. Hillis, C. Moritz, and B. K. Mable). Sinauer Associates, Sunderland, Mass.

Hood, S. B., A. Reznick, and J. D. Palmer. 1994. Phylogenetic relationships in *Anemone* (Ranunculaceae) based on morphology and chloroplast DNA. Systematic Botany 19: 169–200.

Hufford, L., and W. Dickson. 1992. A phylogenetic analysis of Cunoniaceae. Systematic Botany 17: 181–200.

Kesner, M. H. 1994. The impact of morphological variants on a cladistic hypothesis with an example from a myological data set. Systematic Biology 43: 41–57.

Kluge, A. G. 1989. A concern for evidence and a phylogenetic hypothesis of relationships among *Epicrates* (Boidae: Serpentes). Systematic Zoology 38: 7–25.

Kluge, A. G. 1997. Testability and the refutation and corroboration of cladistic hypotheses. Cladistics 13: 81–96.

Kraus, F. 1988. An empirical evaluation of the use of the ontogeny polarization criterion in phylogenetic inference. Systematic Zoology 37: 106–141.

Livezey, B. C. 1986. Phylogeny and historical biogeography of steamer ducks (Anatidae: *Tachyeres*). Systematic Zoology 35: 458–469.

Maddison, W. P., M. J. Donoghue, and D. R. Maddison. 1984. Outgroup analysis and parsimony. Systematic Zoology 33: 83–103.

Nixon, K. C., and J. M. Carpenter. 1993. On outgroups. Cladistics 9: 413–426.

Pimentel, R., and R. Riggins. 1987. The nature of cladistic data. Cladistics 3: 201–209.

Pogue, M., and M. Mickevich. 1990. Character definitions and character-state delimitations: The bête noire of phylogenetic inference. Cladistics 6: 319–361.

Pritchard, P. 1994. Cladism: The great delusion. Herpetological Review 25: 103–111.

Siddall, M. E., and A. G. Kluge. 1997. Probabilism and phylogenetic inference. Cladistics 13: 313–336.

Slowinski, J. 1995. A phylogenetic analysis of the New World coral snakes (Elapidae: *Leptomicrurus, Micruroides,* and *Micrurus*) based on allozymic and morphological characters. Journal of Herpetology 29: 325–338.

Stevens, P. F. 1991. Character states, morphological variation, and phylogenetic analysis: A review. Systematic Botany 16: 553–583.

Strait, D., M. Moniz, and P. Strait. 1996. Finite mixture coding: A new approach to coding continuous characters. Systematic Biology 45: 67–78.

Swofford, D. L., and G. J. Olsen. 1990. Phylogeny reconstruction. Pages 411–501 *in* Molecular Systematics (ed. D. M. Hillis and C. Moritz). Sinauer Associates, Sunderland, Mass.

Swofford, D. L., G. J. Olsen, P. J. Waddell, and D. M. Hillis. 1996. Phylogeny reconstruction. Pages 407–514 *in* Molecular Systematics, 2nd edition (ed. D. M. Hillis, C. Moritz, and B. Mable). Sinauer Associates, Sunderland, Mass.

Thiele, K. 1993. The holy grail of the perfect character: The cladistic treatment of morphometric data. Cladistics 9: 275–304.

Wiens, J. J. 1993. Phylogenetic systematics of the tree lizards (genus *Urosaurus*). Herpetologica 49: 399–420.

Wiens, J. J. 1995. Polymorphic characters in phylogenetic systematics. Systematic Biology 44: 482–500.

Wiens, J. J. 1998a. The accuracy of methods for coding and sampling higher-level taxa for phylogenetic analysis: A simulation study. Systematic Biology 47: 381–397.

Wiens, J. J. 1998b. Testing phylogenetic methods with tree-congruence: Phylogenetic analysis of polymorphic morphological characters in phrynosomatid lizards. Systematic Biology 47: 411–428.

Wiens, J. J. 1998c. Does adding characters with missing data increase or decrease phylogenetic accuracy? Systematic Biology 47: 625–640.

Wiens, J. J., and M. R. Servedio. 1997. Accuracy of phylogenetic analysis including and excluding polymorphic characters. Systematic Biology 46: 332–345.

Wiley, E. O. 1981. Phylogenetics: The Theory and Practice of Phylogenetic Systematics. John Wiley, New York.

Wilkinson, M. 1995. A comparison of two methods of character construction. Cladistics 11: 297–308.

Zelditch, M. L., W. L. Fink, and D. L. Swiderski. 1995. Morphometrics, homology, and phylogenetics: Quantified characters as synapomorphies. Systematic Biology 44: 179–189.

MIRIAM L. ZELDITCH, DONALD L. SWIDERSKI, AND WILLIAM L. FINK

3

DISCOVERY OF PHYLOGENETIC CHARACTERS IN MORPHOMETRIC DATA

In phylogenetic studies of morphological data, shape characters are usually discovered by informal methods of shape analysis and described in qualitative terms. Most often, shapes are compared visually and judgments are based on the perception that forms are either similar or different. Formal quantitative methods (i.e., methods that involve quantitative measurement) rarely are used in the search for characters. In fact, data discovered by these methods are sometimes dismissed as invalid for phylogenetic studies (e.g., Pimentel and Riggins 1987; Cranston and Humphries 1988). Also, conventional methods of phylogenetic analysis, such as coding and parsimony-based methods, are sometimes rejected as unsuited to morphometric data (e.g., Bookstein 1994; Rohlf 1998). Until recently, there was little explicit debate over this issue. But developments in geometric morphometrics (Bookstein 1991) have stimulated interest in morphometric data and have raised questions (and controversy) about appropriate methods for analyzing this type of data (e.g., Fink and Zelditch 1995; Zelditch et al. 1995, 1998; David and Laurin 1996; Lynch et al. 1996; Naylor 1996; Rohlf et al. 1996; Adams and Rosenberg 1998; Rohlf 1998; Zelditch and Fink 1998).

For phylogenetic systematists, the foremost question is whether morphometric data can be analyzed by the methods normally applied to qualitative shape data. The answer to this question is controversial. Some authors see no particular distinction between qualitative and quantitative data, although they may consider particular types of variables unsuited to phylogenetic inference (e.g., Stevens 1991; Thiele 1993; Fink and Zelditch 1995; Zelditch et al. 1995; Swiderski et al. 1998). In that perspective, the same methods and reasoning would apply equally well to both qualitative and quantitative data. Others, however, have gone so far as to say that morphometrics and systematics are incommensurable (Bookstein 1994). In that view, morphometric data could not serve any purpose in systematic studies. Some authors consider morphometrics useful in systematics but reject the idea of applying conventional phylogenetic methods to morphometric data; they recommend alternatives such as classical phenetics or model-based inference (e.g., Rohlf et al. 1996; Rohlf 1998). Still others criticize the use of particular morphometric variables in conventional phylogenetic inference, without stating what type of data or which methods might be more appropriate (e.g., Lynch et al. 1996; Naylor 1996; Adams and Rosenberg 1998).

Our primary aim herein is to explore the legitimacy of morphometric data for conventional phylogenetic studies. To do this, we focus on what we view as the central question: Are qualitative and quantitative shape data categorically distinct? We pose this question because, in the absence of such a distinction, there is no reason to analyze these data differently. We begin by reviewing and analyzing the published arguments against the use of morphometric data in phylogenetic studies. Because we found no clear answers therein, we next compare the reasoning applied to qualitative and quantitative data, using simple examples and a realistically complex data set.

ARGUMENTS FROM THE LITERATURE

Several arguments in the literature suggest that morphometric data may be problematic when used in phylogenetic studies, but our question is whether this type of data is especially problematic relative to other classes of data, or else problematic in ways that no other class of data is. To answer this question, we review four sets of arguments from the literature. One set explicitly contrasts quantitative data to types of data seen as legitimate for phylogenetic studies. The other three sets focus on quantitative data to the exclusion of qualitative data, requiring us to decide if they imply a contrast between qualitative and quantitative data. Of these last three sets of arguments, two are highly general, in the sense that they make no distinctions among kinds of morphometric variables; the third set raises objections to certain types of morphometric variables rather than to morphometric data as a class.

Arguments Contrasting Morphometric and Phylogenetic Data

In this section we include two very different arguments, both of which propose a basic distinction between morphometric data and (legitimate) phylogenetic data. One argument was made by David and Laurin (1996), who asked, "What part should morphometric data play in phylogenetic studies?" They regarded shape data as unsuitable for phylogenetic systematics, recommending instead that the evolution of shape be inferred by comparing a morphometrically based phenogram to a cladogram inferred from qualitative data. The other argument was made by Pimentel and Riggins (1987), who contended that the concept of (taxic) homology cannot be applied to the vague, abstract data of morphometrics. They recommended using qualitative rather than quantitative shape data in phylogenetic studies.

David and Laurin's (1996) argument is not about morphometric data per se, although it may appear to be so, given their question. Their argument is really about shape data in general. They argued that shape cannot be a source of data for phylogenetic inference because shape develops epigenetically and that traits which develop epigenetically are incapable of yielding independent and discrete characters. We note, however, that this argument applies equally well to both qualitative and quantitative shape data. Thus, their logic would exclude both types of shape data from phylogenetic studies.

Pimentel and Riggins (1987) and Cranston and Humphries (1988) did make a distinction between qualitative and quantitative shape data. They regarded qualitative shape data as a legitimate source of characters for phylogenetic analysis, but rejected morphometric data for this purpose. The primary argument in these two contributions seems to be that the concept of taxic homology cannot be applied to morphometric descriptions because they are abstract and vague. These authors also imply that this concept of homology can be applied to qualitative descriptors of morphology because such descriptions are richer and more concrete. Whether these authors' characterization of morphometric descriptions is valid may depend on the particular morphometric methods used to describe shape and also on the accuracy, level of detail, and completeness of the qualitative descriptions. A good qualitative description is better than a simplistic quantitative one, but a qualitative description is not necessarily better than a quantitative description. Also, neither kind of description is necessarily more or less abstract than the other. In one case, the description is framed in the abstract language of words; in the other case, it is framed in the abstract language of measurements.

The secondary argument that Pimentel and Riggins (1987) put forth for rejecting morphometric data concerns ratios. They contended that the numerator and denominator should each be evaluated separately for homology before conclusions are drawn regarding homology of the ratio. This argument is difficult to decipher because it is unclear whether the authors meant merely that the numerators and denominators should be measures of homologous structures or, instead, that the actual lengths observed in a sample of various taxa should be homologous. Of course it is necessary to compare measures of homologous structures; shape variables would not be considered homologous if they were descriptors of different structures. But the other interpretation, that the values of the lengths themselves should be homologous, makes a more questionable argument. Shapes might be homologous even if the values of individual length measurements are not similar. This would be the case when taxa have similar shapes but differ in body size. Systematists who compare taxa that differ in body size, and hypothesize that shape characters are homologous, are undoubtedly concerned with the homology of the structures but not with the similarity of individual length measurements.

In our view, the arguments made by Pimentel and Riggins (1987) bear directly on the question posed in this chapter, but we see no reason for concluding (as they do) that conventional tests of homology (such as those described by Patterson 1982) cannot be applied to morphometric data.

Arguments Focusing on Continuity of Morphometric Data

Many arguments about morphometric data concern the issue of continuity and the methods for coding data that evince overlap among taxa (e.g., Archie 1985; Felsenstein 1988; Farris 1990; Stevens 1991; Thiele 1993; Garland and Adolph 1994; Swiderski et al. 1998). These are probably the arguments most familiar to phylogenetic systematists. Unfortunately, semantic confusion makes a coherent analysis of continuity-related issues difficult: the word "continuous" can refer to features that overlap in their distribution among taxa, and it can refer to an infinitely divisible scale of measurement. The length of a structure would be continuous, in the first sense, when it has the range of 1–10 mm in one species, and 4–12 mm in another; in this case, the two ranges overlap. But the length of a structure would be continuous, in the second sense of the word, even if the range in one species is 1–10 mm but 12–20 mm in another; the ranges do not overlap but length is measured on a continuously valued scale. Thiele (1993) suggested that the word "overlapping" be used for the first sense and the word "continuous" be reserved for the second sense. We follow his suggested usage.

Much of the debate in the literature concerns the coding of overlapping features. There has been debate about whether such features should be used at all and, if so, how to code them. For example, some systematists, such as Pimentel and Riggins (1987), require that characters be mutually exclusive, which rules out using features that any given taxon can express in more than one defined way. Some require that taxa merely have statistically different average values for a given feature (e.g., Farris 1990). And some prefer various other methods to subdivide continuously varying features into discrete states (e.g., Archie 1985).

Even though this debate about coding is undeniably important, it is secondary to the question we ask herein, because the same arguments could be made whether features are described quantitatively or qualitatively. The arguments about overlapping features are not specific to morphometric data; they concern overlapping ranges—and such overlap is not a peculiarity of quantitative data. Only if authors attempt to justify treating quantitative and qualitative data differently is there any implication of a difference between them. For example, if an author requires qualitative features to be disjunct but allows quantitative features to overlap, this inconsistency implies a distinction between the classes of data. But so long as qualitative and quantitative features are treated consistently, there is no implication that they fundamentally differ.

To a lesser extent, debate has also focused on coding features measured on a continuous scale. In this debate, the question is whether a continuous measurement scale should be subdivided at all, even if ranges do not overlap. Some authors object to coding continuously valued variables because it involves subdividing a continuum of values into discrete states (e.g., Felsenstein 1988; Garland and Adolph 1994). Those who oppose coding favor inferential methods that do not require it, such as squared-change parsimony (Maddison 1991) or maximum likelihood (Felsenstein 1981, 1988). Obviously, alternatives to conventional parsimony exist and are technically feasible to apply; but systematists faced with the argument that methods other than parsimony should be used for certain types of data may simply discard such data. After all, if a subset of data makes it necessary to use methods that are based on untested models, perhaps it is better to exclude that subset from analysis altogether.

In our view, there is no reason to abandon either coding or conventional parsimony simply on the grounds that measurement scales are continuous. That type of continuity is immaterial: it does not preclude judging novelty and homology, and coding merely formalizes those judgments. Moreover, the idea that continuously valued variables demand special methods of inference conveys the odd implication that language of description dictates the method of phylogenetic inference. After all, every qualitative shape feature can be redescribed in terms of measurements. Stevens (1991) called such features "pseudoqualitative," a term that would apply to all qualitative size and shape features. In our view, it makes little sense to use entirely different methods of phylogenetic inference depending on the language of description.

We do not mean to dismiss the sometimes difficult and important issues discussed in this section, but they are general problems rather than problems peculiar to morphometric data. Neither notion of continuity suggests that there is any fundamental distinction between qualitative and quantitative shape data.

Arguments Mentioning Only Morphometric Data

The arguments discussed in this section focus on morphometric data, without mentioning qualitative data at all. They concentrate on mathematical properties of morphometric data, which obviously make them difficult to relate to qualitative data. However, relevant mathematical properties of morphometric data, such as multidimensionality, can be characteristic of qualitative shape data as well. In reviewing these arguments, we focus on the properties of morphometric data which lead authors to exclude them from phylogenetic studies, and we ask whether these arguments also would exclude qualitative shape data. The two most important arguments were made by Bookstein (1994) and Rohlf (1998). Bookstein argued that the curvature of shape space renders morphometric data incommensurate with systematics, and Rohlf argued that the multidimensionality (and continuity) of the data makes them unsuited for character-based, parsimony-based systematics.

Bookstein (1994) argued that the curvature of shape space precludes applying basic systematic concepts, especially homology, in morphometrics. The curvature of shape space is undeniable; it follows from the definition of shape widely used in geometric morphometrics (the definition in Slice et al. 1996). But the extent of curvature, and the systematic implications of curvature, are not simple matters of definition or geometry. In Bookstein's view, curvature creates a fundamental impasse to any application of the concept of homology in morphometrics because it means that there are no rectangles in shape space. Lacking rectangles, we cannot compare one transformation, say from A to B, to another transformation, say from C to D. The purely geometric argument is straightforward. However, Bookstein's systematic conclusions do not necessarily follow as a logical deduction from pure geometry.

Bookstein's conclusions are based on two crucial premises. One concerns an empirical matter: that the space of systematic data is so highly curved that conventional biometric approximations would fail. The other concerns the meaning of homology; Bookstein assumed a definition of homology according to which systematists seek evidence of homology by comparing transformations from different starting points (i.e., different ancestors). The first assumption would hold if systematists typically compare highly dissimilar forms. Considering that Bookstein regarded ontogenetic shape change to be well within the limits of the biometric approximations, shapes can differ considerably without differing by too much. And Rohlf (1998) suggested that curvature is not normally a serious problem. However, it could be, and the range of shapes amenable to morphometric analysis remains an issue in need of further empirical study.

Bookstein's second assumption, the one regarding the meaning of homology, is difficult to explain because Bookstein's argument is about rectangles, which have no analog in systematic reasoning. Apparently, Bookstein assumed that a systematist would compare a pair of taxa, such as gerbils and hamsters, and describe the transformation that takes gerbils to hamsters, and then compare another pair of taxa, such as mice and rats, and describe the transformation that takes mice to rats. He assumed that next in the search for homologous features systematists would compare the two transformations to each other and would conclude that the features are homologous if their transformations were the same. Bookstein argued that the two transformations are mathematically incomparable and he therefore concludes that the concept of homology cannot be applied in morphometrics. But to our knowledge, no phylogenetic systematist would ever define homology as the sameness of two transformations from two different starting points; that seems to equate homology to the sameness of modifications of different ancestors (e.g., doubling the length of the height relative to its width, of two different taxa).

Rohlf's (1998) arguments do not focus on the curvature of shape space but rather on the multidimensionality and (to a lesser extent) on the continuity of the data. He raised objections to methods that involve extracting individual dimensions and comparing them separately (i.e., character-based systematics), as well as objections to analysis of such data by parsimony-based methods. He objected to these methods of analysis mainly because they depend on the particular dimensions that are chosen for analysis, which cannot be justified a priori as the uniquely correct ones, and thus the results could be an arbitrary function of the selection of dimensions.

Understanding the technical details of this argument requires some understanding of the metric used in parsimony-based studies: the Manhattan distance. Unlike Euclidean distances, Manhattan distances are sensitive to rotation. (To visualize rotation, imagine tilting the walls and ceiling of a room without altering angles among them and without moving any of the furniture on the floor.) Euclidean distances among the points (i.e., the pieces of furniture) will not change due to rotation of the coordinate system, but Manhattan distances can, being measured parallel to the original orientations of the walls and ceiling. Phylogenetic systematists familiar with the Manhattan distance probably think of this metric in terms of steps on a cladogram. In systematics, the Manhattan distance is calculated by computing the number of character changes along each branch of a cladogram, and summing those to calculate the length of a cladogram (Farris 1983). The most parsimonious tree is the one with the minimum number of steps required by the data; that minimum is obtained by minimizing the Manhattan distance. Even though other distances could be minimized instead, those other distances cannot be interpreted as minimizing the number of ad hoc hypotheses of homoplasy required by the data (Farris 1983). The Manhattan distance is thus not an arbitrarily chosen metric; it cannot be replaced by a different one without altering the logic of phylogenetic inference.

Rohlf's argument concerns the effects of rotating morphometric axes on Manhattan distances. These effects depend on the relationship between morphometric axes and characters, because Manhattan distances are not measured along morphometric axes but rather are measured over characters on a cladogram (Zelditch et al. 1998). If systematists simply count the number of morphometric variables on which taxa differ, one by one, then morphometric variables are equated to characters. But if the morphometric variables are combined to form a single character, the Manhattan distance computed over characters will not equal the Manhattan distance computed over morphometric axes.

Whatever credibility the arguments by Rohlf and Bookstein may have, they do not answer the question that we consider to be central to the debate. They do not suggest any fundamental distinction between qualitative and quantitative data. Bookstein's arguments might seem to, but Bookstein's perception of systematic reasoning is questionable; to our knowledge, no systematist uses the concepts or logic presupposed by him. Rohlf's arguments apply to all complex data, whether qualitative or quantitative, because these are all multidimensional (and continuous). It may be difficult to imagine rotating the "space" of qualitative features, but no one doubts that phylogenetic conclusions depend on the characters analyzed. Whether qualitative shape characters are any more uniquely justified than quantitative ones and whether the conclusions based on quantitative data are any more problematic (or arbitrary) than those based on qualitative data are open questions. Unfortunately, the implications for qualitative data have not been discussed explicitly. As a result, although the arguments were put forth to oppose conventional phylogenetic analyses of morphometric data in particular, they actually oppose conventional phylogenetic analyses of complex data in general.

Arguments Opposing Specific Kinds of Morphometric Data

Several objections are directed at specific morphometric variables. Particularly harsh criticisms are leveled at outline-based morphometric methods on the grounds that these methods disregard biological homology (Bookstein et al. 1982; Zelditch et al. 1995). Others are directed at conventional multivariate variables, such as principal components and canonical variates, on the grounds that these describe dimensions of variation (in the case of principal-components analysis) or axes of dissimilarity (in the case of canonical-variates analysis) rather than characters (Zelditch et al. 1995). To our knowledge, no one has countered that either class of variables is suitable for phylogenetic analyses, although some authors consider them useful for other purposes (e.g., Read and Lestrel 1986; Rohlf et al. 1996; McLellan and Endler 1998; Rohlf 1998). Other arguments are directed at partial warps, one of the variables obtained by geometric shape analysis (Bookstein 1989, 1991). We have argued that these are potentially useful in phylogenetic studies (Zelditch et al. 1995) and have used them in a phylogenetic analysis of ontogeny (Fink and Zelditch 1995). In contrast, several authors see serious problems with partial warps, on purely theoretical grounds (Rohlf 1998) or on grounds explored empirically by applying partial warps in phylogenetic studies (Naylor 1996; Adams and Rosenberg 1998).

Arguments Directed at Partial Warps

One argument against using partial warps in phylogenetic studies is that they are biologically meaningless, being nothing but a priori geometric decompositions of landmark data (the calculation of partial warps is discussed later in this chapter). In sum, partial warps are not features that empirically best differentiate taxa, nor are they features that can be explained as units of biological structure or process. Rather, they are geometrically defined constructs, shape variables arranged in order of decreasing spatial scale (from high scale, such as an anteroposterior gradient, to a more localized scale, such as elongation of the snout relative to the eye). In addition, individual partial warps do not describe the overall change in shape and they are not interpretable in biological terms one at a time. Moreover, in the calculation of partial-warp scores, a particular shape must be designated as a reference form; all specimens are compared to it. This last fact is important because comparisons based on different reference forms will generate different variables and the conclusions drawn from them could conceivably differ.

Rohlf (1998) made all these points, and there is no dispute about this characterization of partial warps. The controversy concerns only their use in conventional systematics. Whether these variables should be rejected as theoretically unsuited to phylogenetic analysis is one subject of debate. The primary argument is against the use of parsimony to analyze partial-warp data (for reasons discussed above, in the section on Manhattan distances). The empirical challenge comes from phylogenetic studies done by Naylor (1996) and Adams and Rosenberg (1998), in which they used partial warps.

Naylor's analysis might best be considered semi-empirical, being a study of imaginary shapes. He drew out a series of shapes, representing an ancestral form and several derived forms. He then computed partial-warp scores by comparing the terminal shapes to the ancestral shape, and he coded the scores as different when the scores differed by a pooled standard deviation of 0.5. These characters were then analyzed as unordered using parsimony, and the results were compared to his true tree. Naylor also reconstructed evolutionary transformations in shape by optimizing the unambiguous characters on the true tree and comparing the inferred shape changes to true changes (i.e., those he had manufactured). (Unfortunately, the true transformations are not depicted in his paper.)

Naylor's data implied two equally parsimonious cladograms and a level of homoplasy that he considered surprising. (No homoplasy was intentionally introduced.) Naylor also judged the reconstructed evolutionary transformations to be inaccurate. These results may seem troubling, but many important questions, such as the following, are left open by his analysis: (1) Would the same conclusions be drawn had he used a more conventional approach to coding, such as inspecting the data for overlap and gaps? (2) Would he have found so much homoplasy in his data had he combined the phylogenetically correlated variables into complex characters, thereby obtaining unique combinations of partial warps? (3) Would his reconstructed pattern of shape evolution be more accurate had he recognized similarities among shapes instead of emphasizing differences (which resulted in many ambiguous characters being omitted from the reconstruction)? (4) Are his expectations reasonable or would most data fail to meet them?

We cannot answer all these questions—only the second and perhaps also the last. It seems clear that he would have found no homoplasy in his partial-warp data had he combined the phylogenetically correlated partial warps into a single character. In his analysis, all the partial warps diagnosing a single taxon are phylogenetically correlated (because all were produced simultaneously by the same strokes of the pen). Thus, all the partial warps optimized to the same branch should be combined into a single character. If that is done, each taxon is diagnosed by a unique combination of partial-warp scores. Then there is no homoplasy in his character data. Not surprisingly, individual components of complex characters can exhibit homoplasy even if the complex, taken as a whole, does not. Thus, Naylor's perception of a surprising contrast between expectations and findings may be questioned because the expectations are based on consideration of the complex, taken as a whole, but the findings are based on individual components. Had he based his expectations on the anticipated behavior of individual components, he might have expected to find homoplasy. Alternatively, had his findings been based on the complex of phylogenetically correlated characters, he would have obtained results consistent with his expectations.

Adams and Rosenberg (1998) focused on related issues raised by Rohlf: the sensitivity of results to the selection of reference form and to the rotation of morphometric axes. The first issue is important to phylogenetic systematists because several different forms may be equally defensible candidates for a reference (such as when studies use multiple outgroups and must choose one to serve as a reference form). The second question is less important because phylogenetic systematists are unlikely to rotate the data arbitrarily (i.e., replace partial warps by principal components of partial warps).

The question concerning the choice of reference arises because partial warps must be based on a single reference form, even if the systematic analysis includes multiple outgroups. Thus, in studies that use an outgroup as a reference form, it is necessary to choose one of the outgroups as a reference form. It is reasonable to ask whether that choice matters to the phylogenetic conclusions. Adams and Rosenberg (1998) concluded that it does, and that it dramatically affects the conclusions. (They inferred two different topologies from their four analyses.) But Adams and Rosenberg changed their outgroup as well as their reference form in each analysis. The obvious question is whether they would have obtained different results had they varied only the reference form, not the outgroup as well. After all, we already know that the choice of outgroup can affect conclusions (Wiley 1981; Donoghue and Cantino 1984; Maddison et al. 1984). A less obvious question is whether their shape data include sufficient information to resolve relationships among the taxa; it would not be surprising to find that weakly supported conclusions are not robust to subtle changes in the data.

In our own recent analysis (Zelditch et al. n.d.), we examined the effects of different reference forms on phylogenetic conclusions. We obtained the same phylogenetic conclusions, regardless of the choice of reference form, under two conditions: (1) when shape data were analyzed in conjunction with other characters, and (2) when the relationships were strongly supported by shape data. When the shape data provide little support for a hypothesis, and they alone are included in the analysis, the choice of reference form matters. Even so, the cladograms based on different reference forms differ by no more than one step from the cladogram inferred from all the data. In addition, we find that different reference forms can (but need not) imply different interpretations of shape evolution. From our findings, we conclude that systematists should either choose a reference based on biological judgment or else use several references and compare results, treating conclusions as only tentative when they depend on a particular choice.

The Status of the Debate: Our Conclusions

Of the arguments reviewed herein, only those made by Pimentel and Riggins (1987) suggest that there is any fundamental difference between qualitative and quantitative shape data. The other arguments either concern generally problematic features of data, such as overlapping distributions, epigenetic development, and multidimensionality, or focus on particular variables, such as outline-based methods or partial warps. In effect, the arguments are either too general or too specific to justify treating quantitative shape data differently than qualitative shape data. Our failure to find rational grounds for distinguishing between qualitative and quantitative shape data could result from the lack of serious efforts to find them. But that failure could have a very different explanation: perhaps the two kinds of data are not really categorically distinct. In the remainder of this chapter, we examine that possibility, by comparing the reasoning applied to qualitative data with the reasoning applied to quantitative data.

ARE QUALITATIVE AND QUANTITATIVE DATA CATEGORICALLY DISTINCT?

We begin our comparison of reasoning with some very simple examples. Our first aim is to articulate the reasoning used in studies of qualitative shape; we then compare that to the reasoning used in quantitative studies of shape. The initial examples, although unrealistically simple, reveal the basic elements of systematic reasoning. We then progress to more realistically complex examples, using data from our own empirical studies. We focus on the reasoning involved in the discovery of characters, because the only role of morphometrics in phylogenetic studies is to aid in the discovery of characters.

The Logic of Character Analysis: Simplistic Examples

Qualitative comparison of triangles. Every systematic study begins by seeking evidence of similarities and dissimilarities. In the case of the simple forms shown in figure 3.1, we begin by sorting them into morphs. This sorting makes one fundamental assumption: that we know which sides (or angles) of the triangle are comparable. If we know which points correspond to each other from triangle to triangle, we can put all of the triangles in a common orientation. Given the assumption that our triangles are all oriented similarly, there is one obvious difference between the two morphs: one is taller relative to its length. We can label that feature as a displacement of the topmost point relative to the bottom side, or we can label it as a change in aspect ratio of the triangle.

Having finished the description, we must code the feature. Also, if we want to say whether the displacement of the topmost point is downward or upward (or if the aspect ratio increases or decreases) we must polarize the feature. Cod-

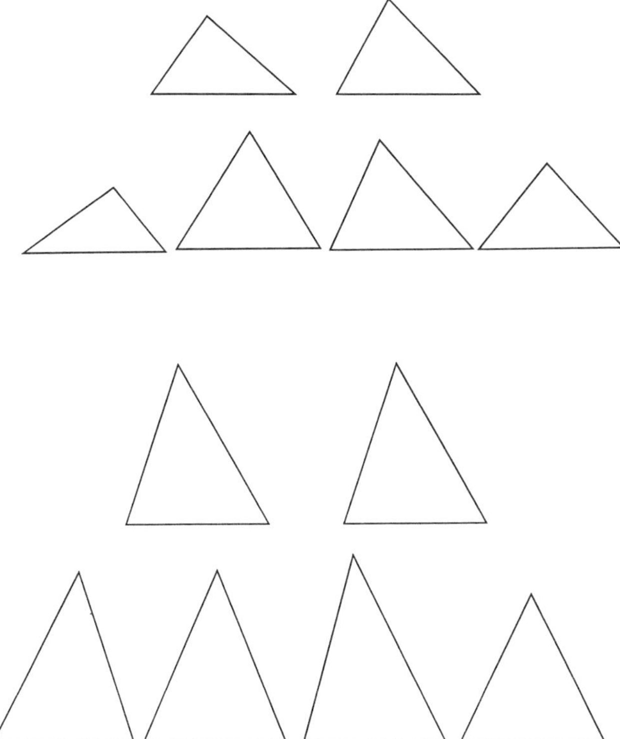

Figure 3.1. A collection of triangles sorted into short and tall morphs.

ing specifies two hypotheses: (1) one of these two shapes is novel (if the character is polarized a priori, that polarity is also specified by coding); and (2) the novel feature shared by two or more taxa is homologous. The judgments of novelty and homology require the reasoning of systematics, not of morphometrics. Once we begin to talk in terms of coding and polarizing characters, we have stopped the morphological analysis and begun a systematic analysis.

Given the simple feature shown in this case, and the information that the outgroup is in the collection of shorter triangles, we would judge the smaller ratio of height to length to be primitive. We would thus propose that the shared derived condition (the larger ratio of the taller triangles) is homologous.

Figure 3.2 also shows a simple case of two morphs, but now they differ by a horizontal shearing of the triangle as well as by a change in aspect ratio. Sorting them is still a straightforward process, but now a question arises: Does this morphological difference imply one character or two? The answer may seem

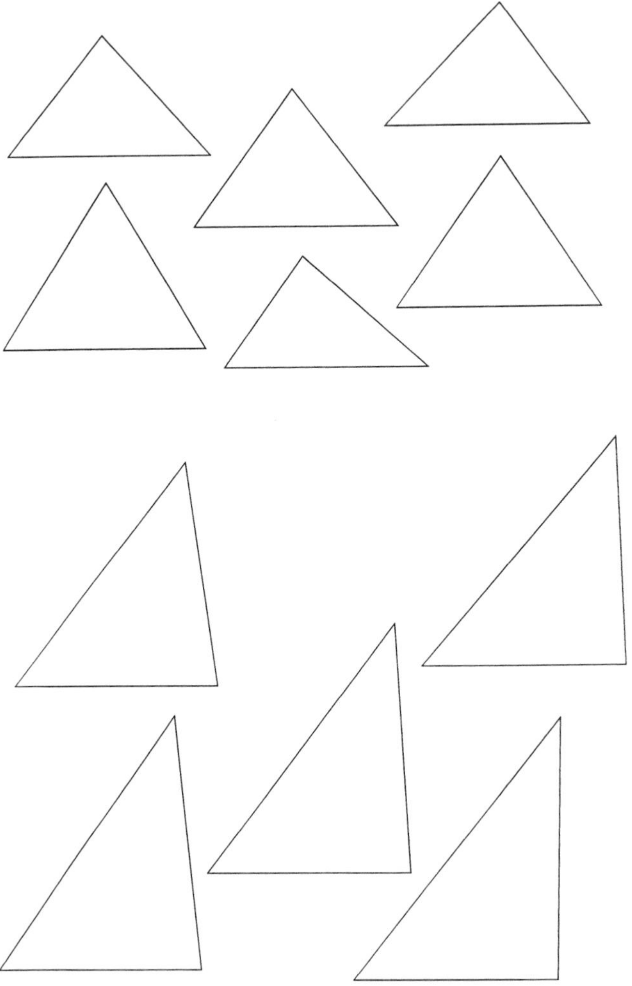

Figure 3.2. A collection of triangles sorted into short and tall sheared morphs.

obvious because every tall triangle also exhibits the horizontal shear. Both features imply the same subsets of taxa. Based on that reasoning, the two features should be combined into a single character. But that reasoning is not the reasoning of comparative anatomy; the decision to interpret the two variables as one character is based on principles of systematics. According to that reasoning, the two features do not seem to be phylogenetically independent. Were we to count them as two features, we would be postulating two historically distinct evolutionary novelties. That seems unlikely given that they have the same distribution in taxa. Of course, they could have the same distribution because they reflect the same history.

There are costs in deciding that these features represent a single character and there are also costs in deciding that these features represent two characters (see Wilkinson 1995 for another discussion of these issues). One cost of combining both features into a single character is that we could underestimate support for a phylogenetic hypothesis. However, there is a cost in treating them separately because, if they are intrinsically correlated, we would inflate the support for the phylogenetic hypothesis by counting that one character twice. Miscounting may not be problematic, but it can be when the two features represent one single homoplastic novelty. Then, by counting it twice, we are introducing correlated homoplasies into the analysis. Thus, the holistic approach (which treats both dimensions as a single character) risks underestimating support for a phylogenetic hypothesis, and the reductionist approach (which treats each dimension as a single character) risks inflating evidence of support and the introduction of correlated homoplasy.

Certainly the accounting of characters matters to phylogenetic conclusions. But such accounting is a distinctly systematic activity. Judgments cannot be based solely on principles of anatomy, but anatomy can provide grounds for deciding among alternative possibilities. For example, anatomy might suggest that two separately described features are really aspects of the same feature because both features of the original description pertain to the same anatomical region (e.g., the dorsal fin of a fish). Development also can provide grounds for deciding among alternative possibilities (e.g., by suggesting that the two features of the original description are developmentally integrated). Thus, both anatomy and development can lead us to question the independence of features. Still, neither anatomy nor development can fully answer the question, because two features drawn from a single anatomical complex can evolve independently and two developmentally integrated characters also can evolve independently. It is the *phylogenetic* correlation of the features that arouses concern for their evolutionary independence, and phylogenetic correlations are not matters of anatomy or development. The interpretation of characters as independent (or not) relies on the evidence of phylogenetic correlations and the reasoning of systematics.

In simple cases like the example above, the reasoning may seem quite straightforward. But the accounting becomes more complex when characters diagnose different (even if nested) subsets of taxa. In figure 3.3, two of the three morphs (B and C) are similar with respect to the vertical position of the topmost vertex of the triangle, but a different pair of morphs (A and B) are similar with respect to the horizontal position of that vertex. If the shorter morph is primitive,

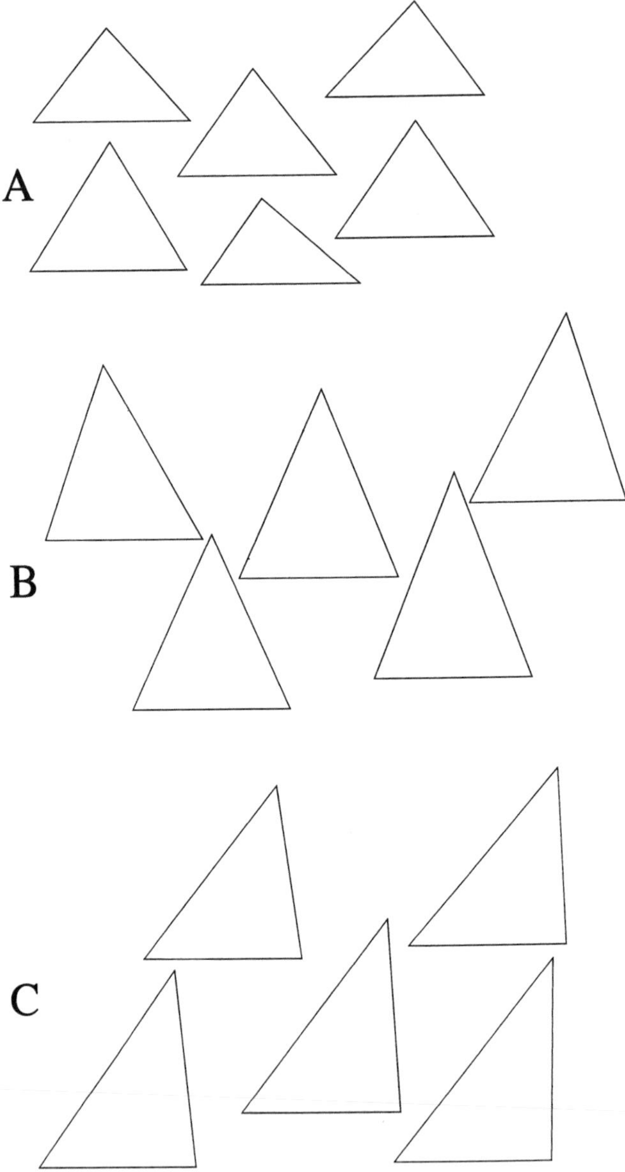

Figure 3.3. A collection of triangles sorted into three morph categories: **(A)** short, **(B)** tall, and **(C)** tall sheared.

as in the previous examples, then at least two different approaches to coding these data could be taken. One approach is to consider them as three distinct morphs and code their shape as a multistate character with three states. Alternatively, the two taller morphs (B and C) might be considered similar and derived with respect to the vertical position of the upper vertex and would be coded as having a shared novelty, and the one that is unique in the horizontal position of that vertex (C) would be coded as having a unique novelty. The latter approach results in two binary characters. The first approach ignores any similarities that exist between morphs but has the advantage of imposing a strict criterion of similarity on all of them. The cost of that strictness is a loss of phylogenetic information: in effect, the three unordered states provide no information whatsoever about relationships among the morphs. Coding each of the morphs as unique would be reasonable if their differences make us hesitant to presume that their similarity is homologous. The second approach extracts more phylogenetic information from the shapes, but at the cost of relaxing the test of detailed similarity. Relaxing the criterion of similarity may make the hypotheses of homology more tenuous.

In figure 3.3, the shape features (vertical and horizontal position of the apical point) do not look to be intrinsically correlated in these taxa. On those grounds, it may seem reasonable to postulate independence of the shape features and thus to treat them as two distinct characters. Still, they could be intrinsically correlated in one taxon and decoupled in another. Character correlations need not be invariant; like morphological features, these correlations can evolve. Although the distribution of features suggests that they are independent, their independence is still a matter of inference rather than observation. The observations, in and of themselves, do not tell us which features are novel nor how those novelties are related historically. Such interpretations follow from the phylogenetic analysis.

Quantitative comparisons of triangles. We can analyze the same three examples with the technique known as "two-point registration" (Bookstein 1991). According to this method, two points on the triangle are assigned coordinates (0, 0) and (1, 0). The same two points are assigned those coordinates on all forms. All of the information about the similarity and difference in shape is contained in the coordinates of the third point, which are called shape coordinates (figure 3.4). The two morphs shown in figure 3.1 can be compared graphically by putting all on the same baseline, and drawing a vector extending between the sets of apical points. Alternatively (and equivalently), the forms can be compared by means of a scatter plot of the shape coordinates (figure 3.5). The axes

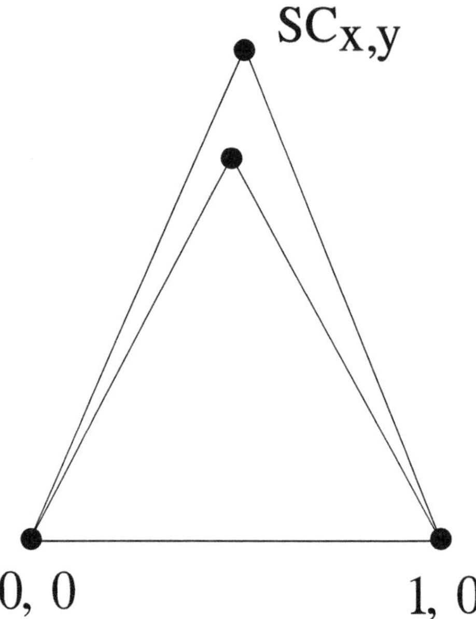

Figure 3.4. Graphical illustration of shape coordinates. Two triangles are superimposed on a common baseline, so that the difference between them is described by the difference between the coordinates of the third point (labeled $SC_{x,y}$).

of the scatter plot are dictated by the baseline: the *x*-axis lies along the baseline and the *y*-axis is perpendicular to it. Both diagrams show that one morph differs from the other in the height of the triangle relative to the baseline—that is, the morphs differ in aspect ratio—which is exactly the same conclusion we came to in the qualitative analysis. And we would code and polarize that feature just as we did before.

In the case of the two morphs shown in figure 3.2, we can again draw out the triangles, or scatter the coordinates, and then draw a vector extending from one morph to the other (figure 3.6). In this case, the orientation or the vector is oblique to the baseline. Given the axes, the feature is two-dimensional. Whether it is a single two-dimensional character or two one-dimensional characters is an open question, and it is the same question we raised in the context of the qualitative analysis. The quantitative analysis does introduce one minor complication: the dimensions are determined by the baseline, so they could change by rotating the baseline. For example, were the baseline oriented in the same direction as the vector extending between the two morphs, then there would be

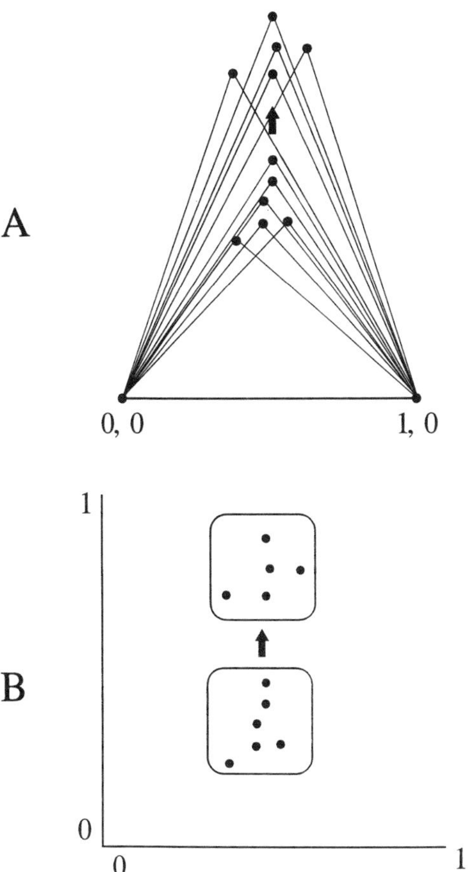

Figure 3.5. Comparison of shape coordinates for the triangles in figure 3.1. The arrow in each panel indicates the inferred direction of transformation between morphs, based on the stipulation that the outgroup is among the short morphs. **(A)** All triangles superimposed on a baseline. **(B)** Scatter plot of the shape coordinates (i.e., coordinates of the upper point of each triangle).

only one dimension of change—along that vector. Obviously, we cannot count morphometric dimensions as if they were real and biologically based, although we could align the baseline with a body axis—e.g., along the anteroposterior axis of a fish. Then we could refer to the dimensions of change in terms of directions on the organism. Nevertheless, even if it appears that the difference lies

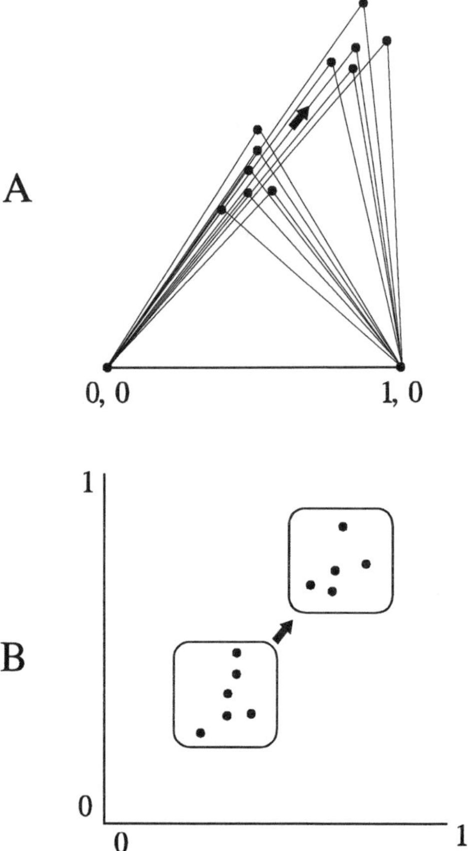

Figure 3.6. Comparison of shape coordinates for the triangles in figure 3.2. The arrow in each panel indicates the inferred direction of transformation between morphs, based on the stipulation that the outgroup is among the short morphs. **(A)** All triangles superimposed on a baseline. **(B)** Scatter plot of the shape coordinates (i.e., coordinates of the upper point of each triangle).

along two perpendicular organismal directions, we must still decide whether that implies a single two-dimensional novelty or two one-dimensional novelties.

In the case of the three morphs shown in figure 3.3, analysis of the shape coordinates reveals that we need two vectors to connect the three morphs (figure 3.7). If we stipulate, as before, that the outgroup has the shorter morphology,

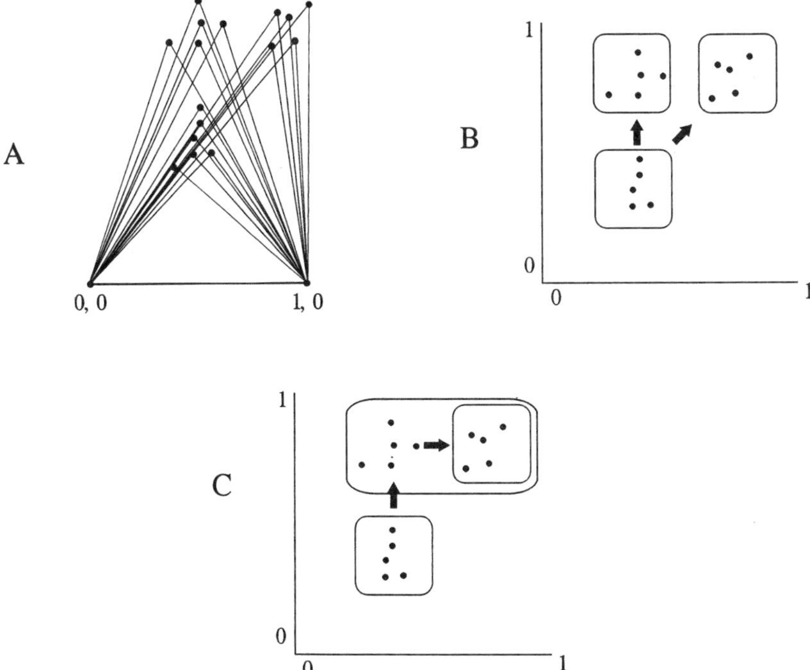

Figure 3.7. Comparison of shape coordinates for the triangles in figure 3.3. The arrow in each panel indicates the inferred direction of transformation between morphs, based on the stipulation that the outgroup is among the short morphs. **(A)** All triangles superimposed on a baseline. **(B)** Scatter plot of the shape coordinates for each triangle, where the boxes around each morph reflect the decision to code triangle shape as an unordered multistate character. **(C)** Scatter plot of the shape coordinates for each triangle, where the boxes around each morph reflect the decision to code triangle shape as two nested two-state characters.

we can draw the vectors in two different ways. One picture implies a two-dimensional multistate character with three states and two directions of change (figure 3.7B). The other picture implies two binary characters, each one-dimensional (figure 3.7C). These are the same alternatives that arose in our analysis of the qualitative data. Quantification allowed us to check our subjective visual impression that the change involved either an increase in aspect ratio or a more complex change in two dimensions. It also allowed us to determine that the two morphs do not overlap. However, quantification did not answer the more basic question: How many characters do we have?

For the example shown in figure 3.7, we prefer the coding scheme that reduces the two-dimensional morphometric variable into two one-dimensional characters. As we discussed above, that decision has a cost, and we cannot know, a priori, whether characters are independent. The advantage of this reductive approach is that we can check the assumption of independence empirically by a posteriori tests (such as the ones we use in our analysis of real data; see the following section). It is more difficult to examine the composite, complex characters to decide whether they should be decomposed. Of course, even in taking the reductionist approach, we cannot avoid the analysis of independence. If we analyze them separately, and do not test characters for independence, we are not actually reserving judgment on this matter. Instead, we would be assuming, or even insisting, that the features are independent. That it becomes necessary to run such tests is the cost of the reductionist approach, but a potential loss of information (contained in the individual dimensions) is the cost of the more synthetic approach.

The problems we discussed above are familiar ones. They are hardly specific to morphometric data; they are not even specific to shape data. Systematists confront the same issues whether analyzing complex morphological features (e.g., eye musculature and the composition of the hypoglossal nerves—Wake 1994) or molecular data (Myers et al. 1995). These authors, and others (e.g., Pogue and Mickevich 1980; Wilkinson 1995; Emerson and Hastings 1998), have raised the same kind of arguments in the broader context of complex systematic data.

Comparisons of Biological Shapes: A Realistic Example

The simple examples reveal some complexities of systematic reasoning. Systematic analysis becomes even more complicated when the shapes are complex and when they vary and undergo ontogenetic change. To examine distinctions between qualitative and quantitative data further, we now turn to more realistic examples.

Complexities of real data. In our studies of piranha shape, we analyze 16 landmarks on the body of the fish (figure 3.8). As figure 3.9 shows, even if we analyze only two triangles (i.e., four landmarks), we would already face more serious questions about independence than we did in the case of single triangles. In figure 3.9A, we have drawn four landmarks and two triangles on a fish; the second triangle overlaps most of the first. The question is whether the two triangles are independent. Independence seems implausible given their anatomical positions. In figure 3.9B, the second triangle is ventral to the first and they do not overlap. Nevertheless, we face the same question because both triangles contain the midbody of the fish.

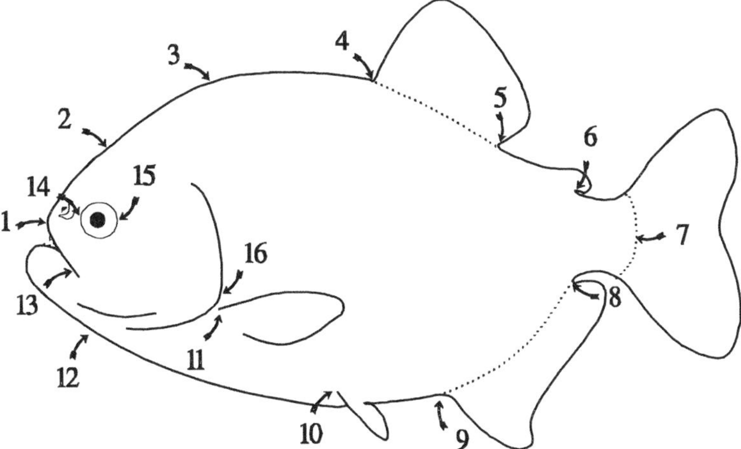

Figure 3.8. Locations of the 16 landmarks used in the morphometric analyses of piranhas.

Our analysis of shape, based on 16 landmarks, is clearly an analysis of multidimensional data. Because there are 16 landmarks, the shapes we are analyzing exist in a 28-dimensional space. It is not self-evident which dimensions correspond to independent characters. Of course, that is not self-evident before we quantify our comparisons, either. Qualitative shape data, unless reducible to a single ratio of a single triangle, are necessarily multidimensional. Quantification does not exacerbate this problem, it merely provides a language for articulating the problems posed by it.

Multidimensionality is not the only problem we face in studies of real data. Unlike the simple triangles in our previous examples, real organisms do not necessarily fall into discrete clusters of shapes. Individuals vary within populations and sometimes the ranges of taxa overlap so broadly that populations are morphologically indistinct. In addition, shape is dynamic through the lifetime of a single individual. Both individual variation and ontogenetic change can complicate the search for evolutionary novelties. In figure 3.10 we show two comparisons of a single shape feature in juveniles and adults of six species of piranhas. The juveniles are easily sorted into two morphs; the two species of *Serrasalmus* have shallower bodies than do those of *Pygopristis* and the three species of *Pygocentrus*. In contrast, the adult morphs are not so easily delimited.

The pattern shown in figure 3.10 for the adult morphs is the one typically ascribed to quantitative data. It receives so much attention that it may be responsible for much of the opposition to morphometric data. But it is the organ-

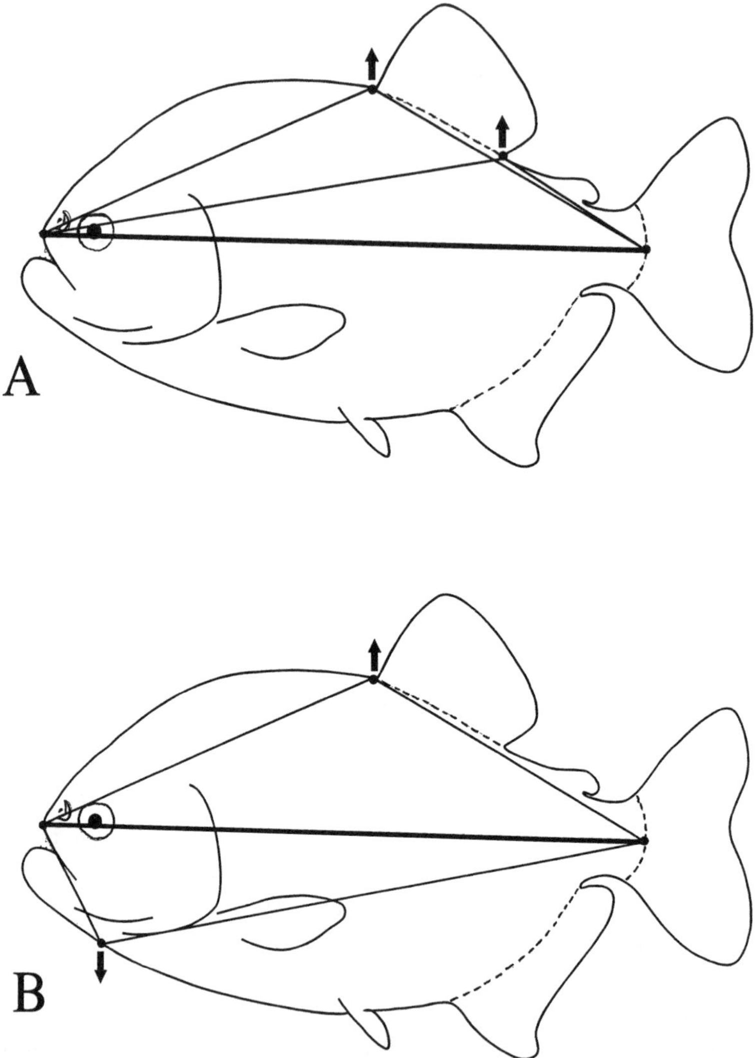

Figure 3.9. Two triangles drawn on a fish. **(A)** Virtually overlapping triangles. **(B)** Dorsal and ventral triangles that share only a common baseline but include portions of the midbody.

isms that generate this pattern, not quantification of their features. Had we attempted to analyze fin positions qualitatively, we would have discerned the same overlap. The problem in such cases is that the organisms (with the exception of *Serrasalmus elongatus*) do not reveal evidence of an evolutionary

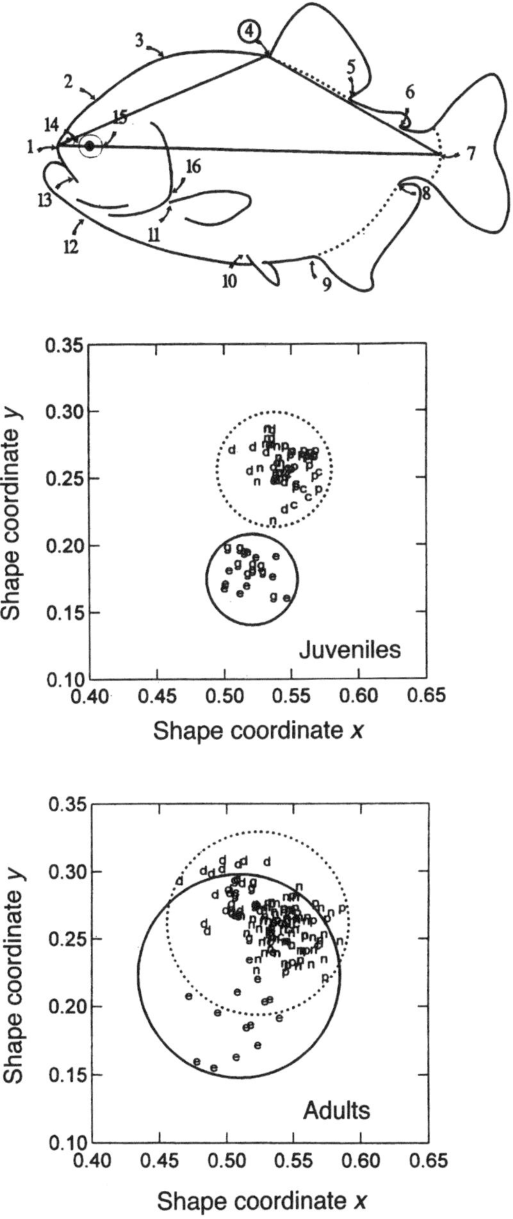

Figure 3.10. Comparison of shape coordinates (at landmark 4) for one triangle in six species of piranhas at two ages. Symbols in the scatter plots denote piranha species (d, *Pygopristis denticulata;* c, *Pygocentrus cariba;* n, *Pygocentrus nattereri;* p, *Pygocentrus piraya;* e, *Serrasalmus elongatus;* g, *Serrasalmus gouldingi*).

novelty. With the one exception, they are too similar to each other to designate any feature as distinct. That lack of evidence of novelty is hardly the fault of quantification; quantification merely reveals that the ranges of the taxa overlap. And, as shown in figure 3.10, the extent of overlap may be age-specific. The comparison of juvenile morphologies suggests that there is an evolutionary novelty. And the comparisons of the two pictures of static shape suggest that there may be informative features in the ontogenies of these shapes.

Advantages of morphometric analysis. The complexities of real biological data, especially the complexities introduced by individual variation, overlapping distributions, and ontogenetic change, provide the primary rationales for morphometric analyses. Morphometrics supplies techniques for explicit and objective analyses of variation and overlap, and it also supplies methods for describing ontogenetic transformations. There are two additional and important advantages in morphometric studies: the descriptions of features are often more precise than in qualitative studies, moreover, morphometrics includes tools for analyzing spatial covariances among landmarks.

The advantage of precise description is hard to overstate. For example, in studies of piranhas, body depth is often compared among species (e.g., Eigenmann 1915; Norman 1929; Fink 1993). The meaning of the term "body depth" can be quite vague as it could refer to a general summary of depth over the whole body (the body, as a whole, is generally deeper), or it could refer to a depth of a particular anatomical region. A quantitative analysis forces us to specify exactly where body depth is measured and exactly how evidence of a difference in depth is judged. Moreover, the scatter plots make the evidence accessible to the reader. For example, if we were to show the scatter plot of the adults' data from figure 3.10 and claim that it reveals evidence of a difference in body depth, readers would be able to examine the pattern for themselves and judge whether our conclusion were justified. Instead of creating ambiguity, the morphometric analysis reveals it.

The advantages of morphometric analysis are particularly clear when we seek to analyze ontogenetic data. Systematists have long been urged to compare whole ontogenies (Danser 1950; Moser et al. 1984; de Queiroz 1985; Kluge and Strauss 1985; Creighton and Strauss 1986). Ontogenetic comparisons are useful for two reasons: (1) there may be more phylogenetic information in whole life-cycles than in single discrete stages, and (2) there may be evidence of character independence in the lack of ontogenetic correlations among features. Figure 3.11 shows an analysis of the ontogenetic trajectories of two features of piranha shape. The two features are the relative heights of the anterior and posterior ends of the dorsal fin. The ontogenetic analysis involves familiar

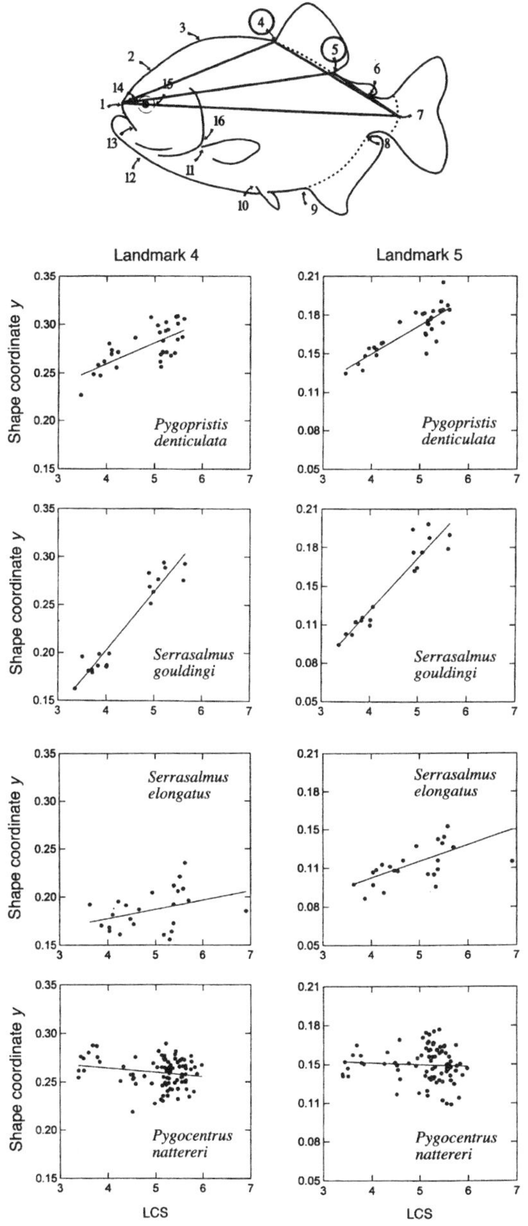

Figure 3.11. Comparisons of ontogenetic trajectories for the *y*-axis shape coordinate at two landmarks (4 and 5) for four piranha species. *Pygopristis denticulata* is the outgroup. LCS = log centroid size.

techniques of regression because we are simply analyzing the ontogenetic association between size and shape. As the figure shows, the two landmarks show similar relationships between relative height and body size. That is hardly surprising both because these points are at vertices of overlapping triangles and because the two *Serrasalmus* species have generally shallow-bodied juveniles, so the starting points of the trajectories are lower for both points in both species.

If we treat the position of each landmark as a single independent feature, then we are treating each triangle as a single character. Just as some morphological features, like body depth, are described by multiple triangles inscribed on the body, there also can be ontogenetic transformations described by changes in multiple triangles (e.g., general shallowing of body depth or anteroposterior growth gradients). Morphometric analysis of ontogeny permits detection of such patterns and, more important, it permits comparisons to determine whether these patterns have undergone phylogenetic transformation.

Analysis of spatial covariances. In the various examples above, we have mentioned spatial covariances as a possible source of phylogenetic correlations. Whereas phylogenetic correlations are a matter of historical pattern, spatial covariances are a matter of anatomical geometry. No morphometric method can tell us which triangles are phylogenetically independent, but one method can tell us about the spatial aspect of the correlations among triangles: the thin-plate spline decomposed by its partial warps (TPS). This method has been described elsewhere (e.g., Bookstein 1989, 1991; Zelditch et al. 1992; Swiderski 1993), but some of the details are important for understanding its value in systematic studies. (Software for conducting the analyses is available at http://life.bio.sunysb.edu/morph.)

In TPS analyses, shape change is modeled as a deformation. Figure 3.12 shows deformations describing ontogenetic transformations in several species of piranhas. A deformation is a smooth mapping of the points in one form to corresponding points in another form. The thin-plate spline is a particular model of a deformation, one based on a physical metaphor: the bending of an idealized steel plate (one having uniform composition, infinite area, and infinitely thin). According to this metaphor, displacements of landmarks in the x,y-plane are visualized as if they were transferred to the z-coordinate. Thus, instead of depicting change at a landmark along the dimensions of the picture plane, the movement is visualized as if it occurred in the third dimension. The initial positions of the landmarks are represented as a set of points on the steel plate. The distances that the landmarks would have to move to reach their new positions in the plane of the data are represented by a set of stalks attached to the initial positions and projecting up or down from the plate. Then, the steel plate is bent

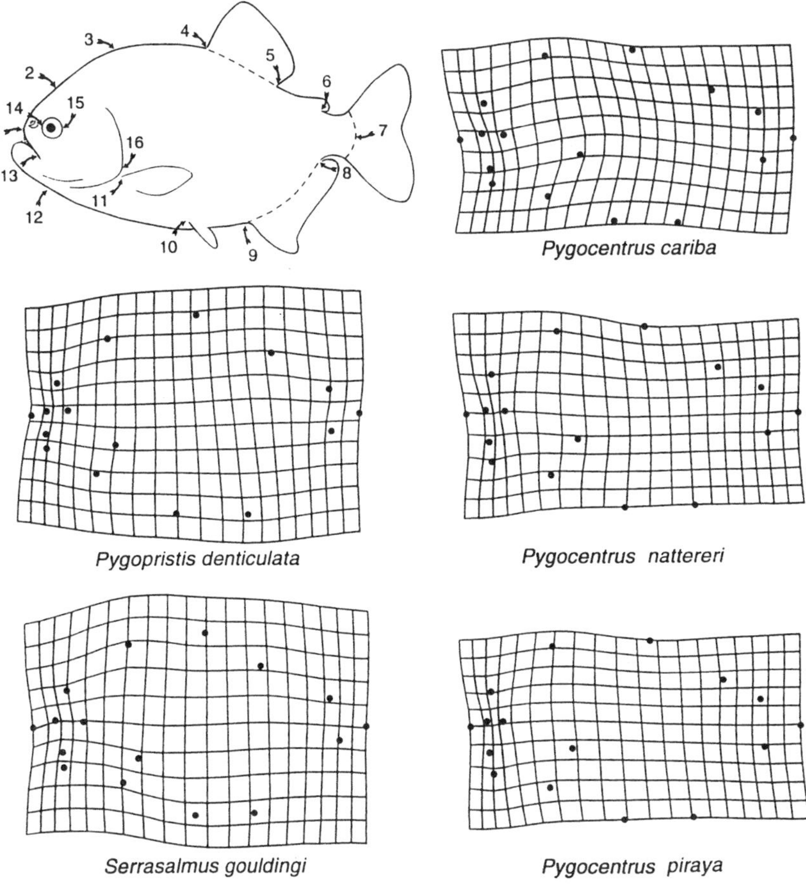

Figure 3.12. Deformations describing ontogenetic transformations of five species of piranhas. *Pygopristis denticulata* is the outgroup.

so that the points at the bases of the stalks are pushed out to the tips. The new conformation of the plate is described by a formula that minimizes the magnitude of bending over the whole plate, and also minimizes the physical energy required to produce that bending.

The energy required to bend the plate depends on the physical distances between landmarks in the original configuration and the relative heights of the stalks representing the displacements of the landmarks to their new positions in the *x,y*-plane of the real data. If two plates are bent by the same amount (so that they describe the same total amount of shape change) the one that is bent between closely spaced points requires more energy. More technically, bending energy is a function of the rate of change in the slope of the plate (i.e., whether the slope of the surface increases rapidly or slowly). The function that is mathematically minimized is the rate of change in slope of the surface. To interpret this bending in terms of changes in shape, we project those changes visualized in the *z*-direction back into the plane of our data (separately for the *x* and *y* coordinates) where they are represented by the deformation of square grid drawn on the first plate.

TPS leads to an analysis of spatial covariances because the idea of bending is interpreted as spatially local information. For example, the displacement of a single landmark relative to the surrounding area could be interpreted as a highly localized deformation affecting only the immediate vicinity. The interpretation would require a large displacement of the landmark in one direction, large displacements of nearby points in the opposite direction and relatively rapid changes in slope which would require high bending energy. A gentler change in slope affecting a larger area with a greater distance between points moving in opposite directions would require less bending energy. Thus, minimizing the bending energy function is equivalent to minimizing spatially localized information. That is useful because it is always possible to interpret any change as local to a single point, but that would be the most unparsimonious interpretation possible. If we construe each individual landmark as if it were displaced independently of all others, we require the steel plate to have many steep-sided peaks and narrow pits and we require the deformation to have a high bending energy. By minimizing this function, changes are interpreted as local only when data require it (i.e., only when we observe a contrasting displacement of a nearby landmark). Thus, minimizing bending energy leads us to a more parsimonious description of the change; spatially correlated changes are interpreted as a single feature.

In some patterns of landmark displacement, the steel plate describing the shape change does not bend at all, it simply tilts or rotates. Such changes are called "uniform" or "affine" and they describe a global change, one that is the same everywhere over the whole organism. In other cases, the plate bends and in those cases the changes are called "nonuniform." These comprise changes in shape that vary from one region to another over the organism. In most cases, including all the data analyzed by us, changes in shape are combinations of both uniform and nonuniform components (figure 3.13). The nonuniform part of the deformation can be decomposed (by eigenanalysis of the bending energy matrix) into a series of nonuniform components at progressively smaller spatial scales. Figure 3.14 shows three components of the nonuniform part of the deformation shown in figure 3.13, in order of decreasing scale (or greater local-

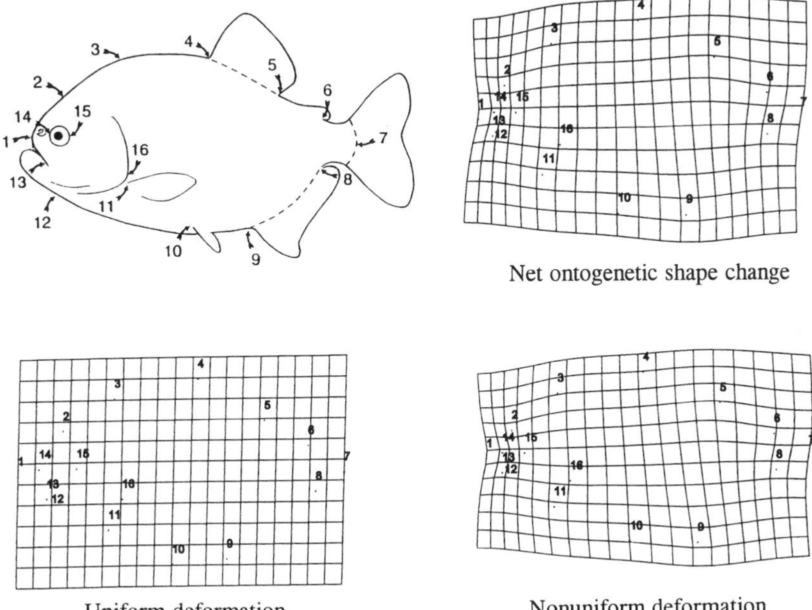

Figure 3.13. Uniform and nonuniform components of the deformation describing the ontogenetic transformation of *Pygopristis denticulata*.

ization). Figure 3.14A shows a change at a large spatial scale, that is, one broadly distributed over virtually the whole organism. When oriented along the anteroposterior body axis, as shown here, it can be interpreted as an elongation of the midbody relative to more cranial and caudal regions. Figure 3.14B shows a feature at a moderate spatial scale, describing a more spatially confined change in the posterior region of the body. As oriented here, the picture shows a deepening and elongation of the region between the dorsal and adipose fins relative to the caudal peduncle. More distant landmarks are not involved in the description of this change, hence it is more localized. Figure 3.14C shows a somewhat more localized feature that is confined to the cranial region.

The number of components and where they localize change is biologically arbitrary. The number of components is determined by the number of landmarks. The specific pattern of localized change described by each component is determined by the spacing of landmarks on one form (called either the reference form or the starting form) to which all others are compared. These components do not indicate where change actually occurs; that is determined sepa-

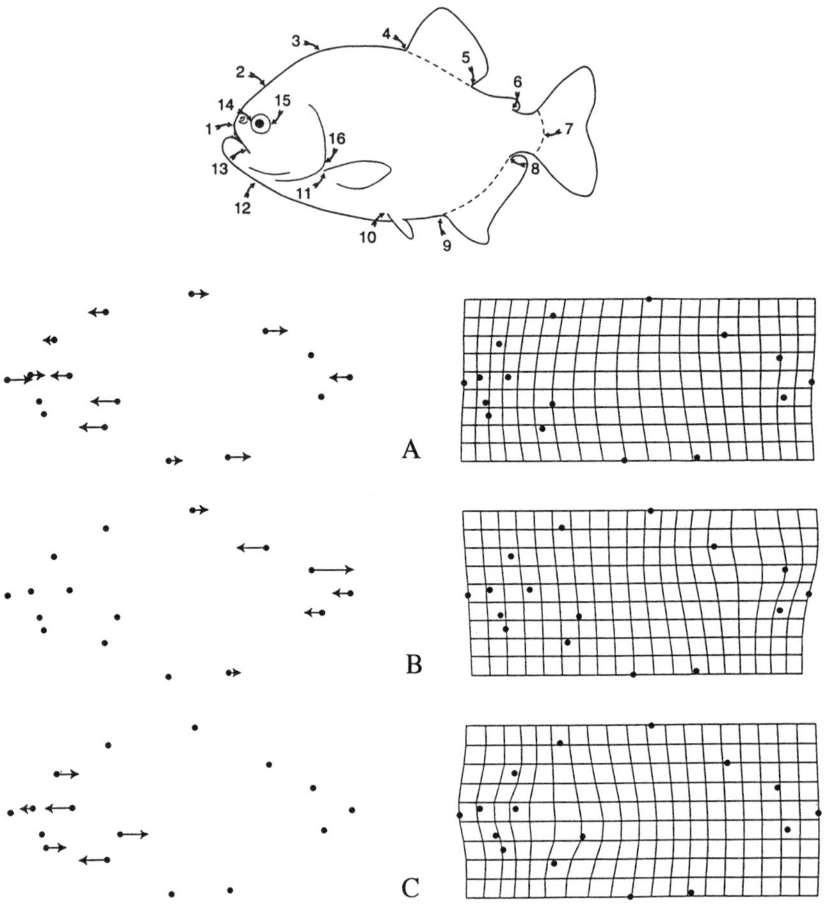

Figure 3.14. Difference in spatial scale of selected of components of the nonuniform component of the deformation shown in figure 3.13. **(A)** A component at large spatial scale. **(B)** A component at moderate spatial scale. **(C)** A component at smaller spatial scale.

rately. Rather, they are simply a set of features available for comparison. Taken together, the set of all components, including the uniform and the complete set of nonuniform components, provides a basis for comparison (a coordinate system of orthogonal axes). This basis differs dramatically from those based on the sample variance-covariance matrix, such as principal components or canonical variates. Because the components of deformation produced by TPS are determined solely by the spacing of homologous landmarks in the starting form and are computed before any empirical analysis begins, they are purely geometric and are not functions of the dimensions of variation in a sample. Thus, these geometric components may seem odd because they are calculated before any comparisons are done, but they are useful for exactly that reason. They are not defined in terms of phenetic distances, but are functions of the anatomical distances among homologous points on the organism.

Given these components, we can compare taxa and ask how they differ. For each comparison made—i.e., from each specimen to the same reference form—we obtain coefficients, called partial-warp scores. The coefficients tell us how much, and in what direction, the shapes differ. The relative magnitude of each score tells us how much that component contributes to overall dissimilarity. Although we are not interested in overall dissimilarity, we are interested in the evidence of novelty of individual features. To be novel, a feature must be different, so we need a concept of dissimilarity (a distance metric) before we can inspect such scores for evidence of novelty.

We also need a description of the direction of change. To explain the concept of direction, we show one component oriented in three directions (figure 3.15). In all three frames, each vector has the same length relative to the others, and the same vectors point toward each other or away. All that differs is the orientation of the arrows (the full complement of arrows in each figure represents a single partial warp). The whole set of arrows can be oriented along the anteroposterior body axis (figure 3.15A), or along the dorsoventral body axis (figure 3.15B), or in some direction between the axes (figure 3.15C).

Partial-warp scores, like the shape coordinates we have examined so far, can be inspected for their scatter and they can be regressed on size. Partial-warp scores can be treated exactly like any conventional morphometric variable. There are some fairly specialized techniques for analyzing them because some methods treat shape as a single unitary feature and so adopt an essentially univariate approach to shape (i.e., they regress the variable "shape" on size; the program tpsRegr [Rohlf 1997] performs such analyses). But all of the methods that are conventionally applied to morphometric data can be applied to partial-warp scores. Just as we used regression to analyze the ontogenetic trajectories of shape coordinates, we can use regression to analyze the ontogenetic trajectories of partial-warp scores.

As in the analyses of shape coordinates, we must decide whether to treat changes in both directions of a partial warp (or in several partial warps) as one or more features. Just as we cannot assume that features of triangles evolve independently, we cannot assume that the features described by geometric components of a deformation evolve independently. Several partial warps may describe a single morphological feature, so it is necessary to analyze their

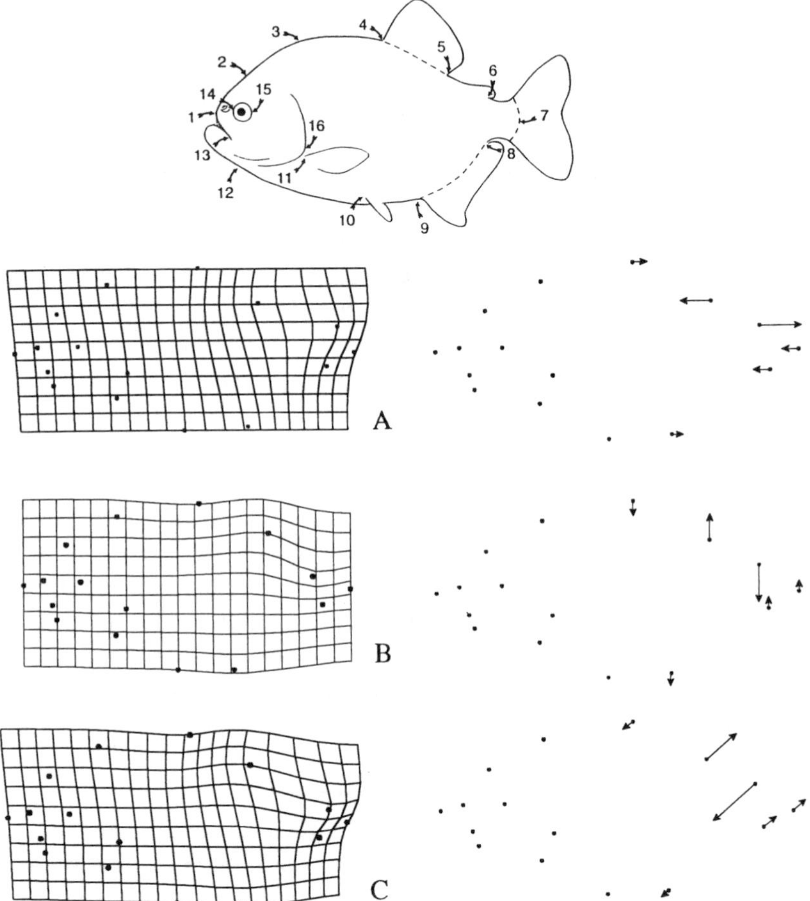

Figure 3.15. A single nonuniform component oriented in three different anatomical directions: **(A)** parallel to the anteroposterior axis of the fish; **(B)** parallel to the dorsoventral axis of the fish; **(C)** at a 45° angle between the anteroposterior and dorsoventral axes.

phylogenetic correlations empirically, as we will show, below. Although TPS provides a tool for analyzing spatial covariances, it does not provide a tool for analyzing phylogenetic covariances. To analyze phylogenetic covariance we must make the transition from variables to characters, and from morphometrics to phylogenetics. Thus we now turn to the second phase of analysis, the systematic phase, in which we assess evidence of novelty and similarity, and code features according to that evidence, and then use the data to infer topologies of taxa and patterns of character evolution.

Systematic Analysis of Morphometric Data

Coding. Once we have descriptors of shape similarities and differences, we can begin the systematic analysis. The step at which we begin to interpret observations in systematic terms is coding. Coding is a complex process, requiring judgments of character individuality, novelty and homology. The function of coding is to specify hypotheses of novelty and homology, which are subsequently tested by parsimony. We have argued that the reasoning used to code quantitative data should be consistent with the reasoning used to code qualitative data (Swiderski et al. 1998). We therefore object to methods specifically tailored to quantitative data when they invoke principles that would be rejected in qualitative studies. Of course, systematists differ on criteria for coding. We (Swiderski et al. 1998) code features as distinct when their ranges do not overlap (or they overlap so slightly that the feature would be construed as a novelty in a qualitative study). However, our basic argument is that the criteria should be applied consistently to qualitative and quantitative data.

Ontogenetic data pose special problems for coding because we usually cannot observe the ontogeny of each individual. For that reason we cannot examine variation of ontogenies and ask if ranges overlap, as we would do in a study of static shape (see Swiderski et al. 1998). Instead, we can only obtain a coefficient representing the ontogenetic change of an average individual. In our studies of ontogenetic data, we determine whether there is (statistically significant) ontogenetic change in a feature, and if so, whether the allometry is positive or negative (Fink and Zelditch 1995). Taxa are coded as similar if they are similar with respect to the presence or absence of allometric change, and then, if such change is present, if they are similar with respect to direction of allometric change.

In our view, quantitative data pose no special problems for coding beyond the problems inherent in the shapes before quantification. In spending so little time on this issue we do not mean to trivialize it, but it is not an issue peculiar to morphometric data.

Using shape coordinate or partial-warp data. We focus here on our ontogenetic data (see Swiderski et al. 1998 for a discussion of static shape data). Having coded ontogenetic transformations of both shape coordinates and partial warps, we can now infer cladograms implied by these data. We can also ask

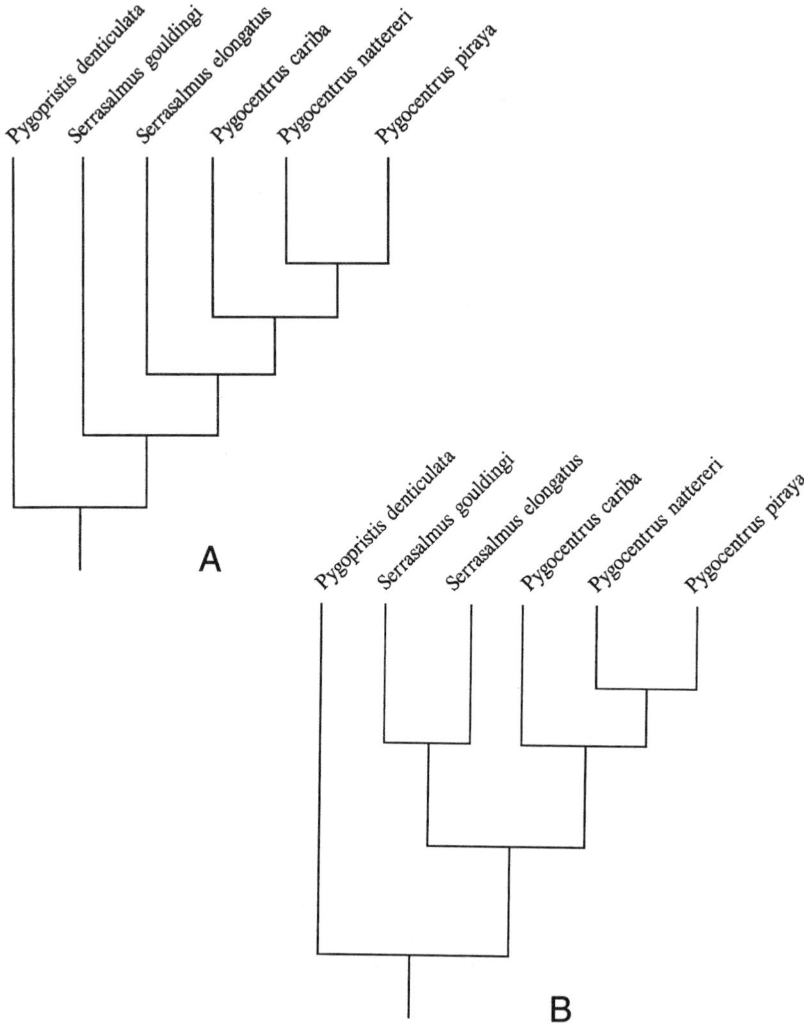

Figure 3.16. Cladograms inferred from two different sets of morphometric data combined with the same set of traditional data. **(A)** Cladogram implied by the data set that includes shape coordinate data. **(B)** Cladogram implied by the data set that includes partial-warp data.

if the different data sets imply the same topology. Because the shape coordinates and partial warps describe exactly the same features, they are not logically independent; and therefore, we must analyze these data sets separately. However, we can (and do) combine our shape data with all our other characters.

For all of the analyses described below, *Pygopristis denticulata* is our outgroup, multistate characters are unordered, and the analyses are done using PAUP 3.1.1 (Swofford 1993). Our preliminary analyses assume that each dimension of each shape coordinate is independent, and also that each dimension of each partial warp is independent. These assumptions are checked a posteriori, and should they prove unrealistic, the phylogenetically correlated variables are combined into a single character, recoded, and reanalyzed. (We discuss methods for judging these assumptions a posteriori in the context of our results.)

Each of the two preliminary data sets implies a single most parsimonious cladogram (figure 3.16). These differ in topology. According to the cladogram implied by the data set that includes shape coordinates, *Serrasalmus* is paraphyletic; *S. elongatus* is the sister group of *Pygocentrus* (figure 3.16A). In contrast, both *Serrasalmus* and *Pygocentrus* are monophyletic according to the cladogram implied by the data set that includes partial-warp data (figure 13.16B, which is the topology implied by all the other data, excluding shape characters). That conflict, however, could conceivably be an artifact of character correlations. Several correlated homoplasies may be responsible for one (or both) patterns. In light of that possibility, we must examine the inferred homoplasies and the inferred homologies for their phylogenetic correlations. At this point, the a posteriori tests of character independence become necessary.

Four shape coordinates (at landmarks 4, 5, 9, and 10), all of which are in the dorsoventral direction (the *y*-direction, given our anteroposterior baseline), support the sister-group relationship between *S. elongatus* and *Pygocentrus* (figure 3.17). But, as evident from figure 3.18, two of these points are adjacent to each other on the dorsal body (at anterior and posterior bases of the dorsal fin) and two others are directly below them, also adjacent to each other. Their anatomical positions provide one reason for doubting their independence. In addition, in *Pygopristis* and *S. gouldingi,* all four move vertically, indicating a deepening of the body (both dorsally and ventrally) in this region. That deepening, in this particular region, does not occur in either *S. elongatus* or *Pygocentrus*. It may even be difficult to find two of the points in the diagram of *Pygocentrus* because there are no arrows extending from them at all—they are simply points.

Based on the anatomical position of these points, as well as a simple ontogenetic explanation for their correlation (a change in the deepening of the posterior body), the hypothesis of four independent novelties seems doubtful. Thus, these points fail tests of independence based on anatomical position and on ontogeny. For those reasons, we recode them as a single character and obtain a different cladogram (figure 3.19). This cladogram implies the same relationships as the one inferred from the data set that included partial-warp scores. However, we cannot conclude yet that shape coordinates and partial warps will lead to the

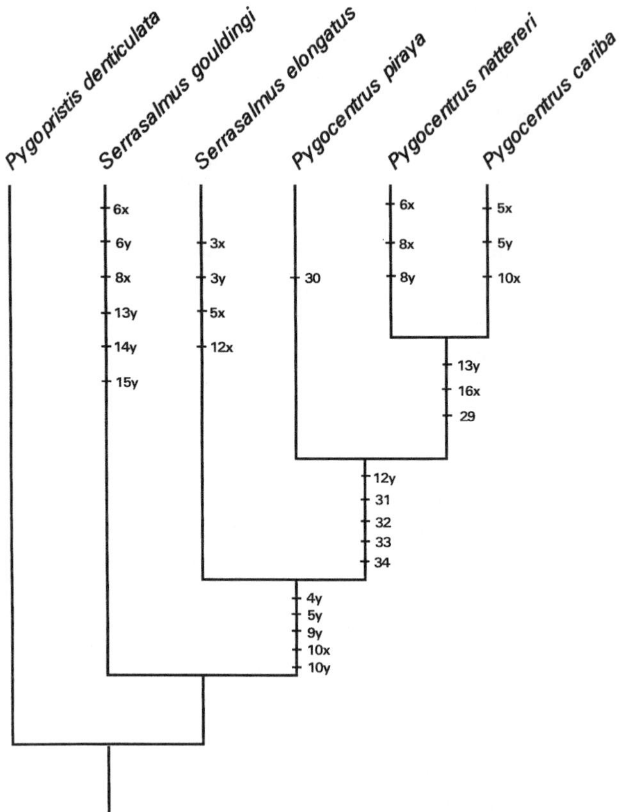

Figure 3.17. Cladogram inferred from shape coordinate data and traditional data, showing the distribution of unambiguously optimized characters. The numbers 1–16 represent the shape coordinates and correspond to the numbers of the landmarks (figure 3.8); the *x* and *y* labels describe the directionality of the changes (along the anteroposterior and dorsoventral axes, respectively). The numbers 29–43 represent traditional characters.

same phylogenetic conclusions. We must examine all the hypotheses supported by the shape coordinates (both hypotheses of homology and homoplasy) and test the independence of each one just as we did for these four. Should we find at the same node that all describe a change in a single region of the body, or that all have a common developmental explanation, we would recode those as a single character and reanalyze the data once again. This process can be tedious, and some decisions will be difficult to make; nonetheless, these tests are necessary.

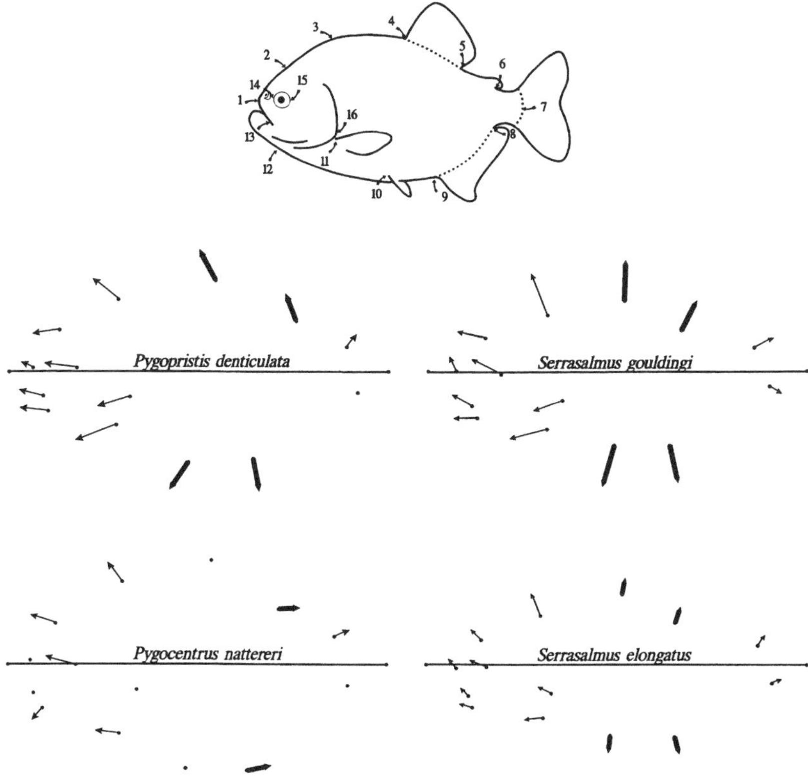

Figure 3.18. Ontogenetic trajectories of shape coordinates for four species of piranhas, including the outgroup *(Pygopristis denticulata)*. Arrows in bold represent features that are inferred to be correlated.

The partial warps must also checked for their phylogenetic correlations. Figure 3.20 shows the distribution of unambiguous partial warps on the tree. Two partial warps (as well as other data) support the hypothesis that *Pygocentrus* is monophyletic. One of these is a feature lost in the ontogeny of *Pygocentrus,* the deepening of the region between the dorsal and adipose fins relative to the caudal peduncle (figure 3.21A). The other describes a feature gained in the ontogeny of *Pygocentrus,* the deepening of a more ventral posterior region relative to the anterior belly and a change in the ventral profile (figure 3.21B). These features may also be correlated because one spatial pattern of deepening may be replaced by the other. If so, this substitution of one feature for another cannot be considered as two independent characters. Reasoning on these grounds, we would conclude that these two features fail a test of ontogenetic independ-

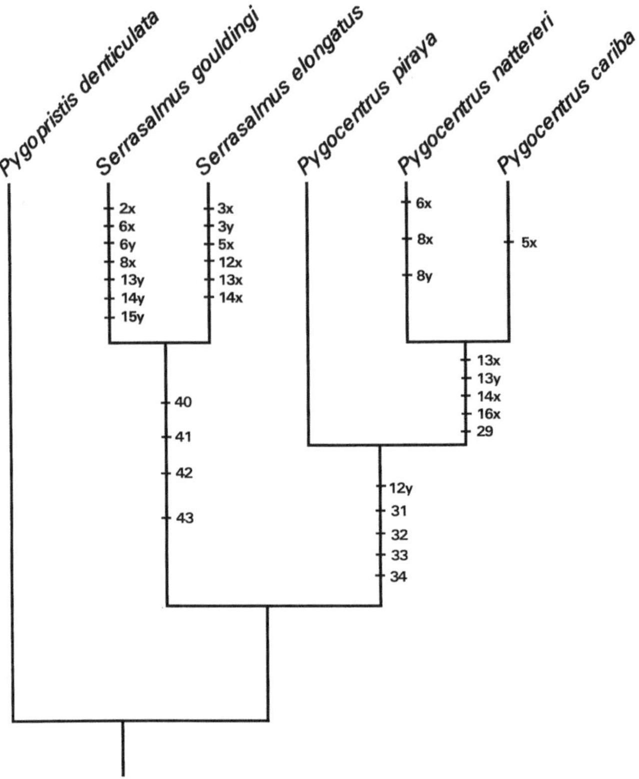

Figure 3.19. Cladogram inferred from shape coordinate data and traditional data after the four features inferred to be correlated were recoded as a single character. Symbols showing the distribution of unambiguously optimized characters are as defined for figure 3.17.

ence and recode them as a single character. Doing so, we obtain the same topology as before, but with one less character supporting the monophyly of *Pygocentrus* (figure 3.22). Of course, we must evaluate the independence of the other partial warps, checking them for anatomical overlap and developmental independence, but it should be clear from this comparison that partial warps and shape coordinates do not necessarily lead to different phylogenetic results.

Unfortunately, we cannot give absolute rules for deciding when hypotheses of character independence can be safely accepted. The tests exemplified here, which are based on phylogenetic pattern (features that diagnose the same group), on anatomical position, and on developmental integration, as well as tests based on features which change together on multiple branches (see Fink

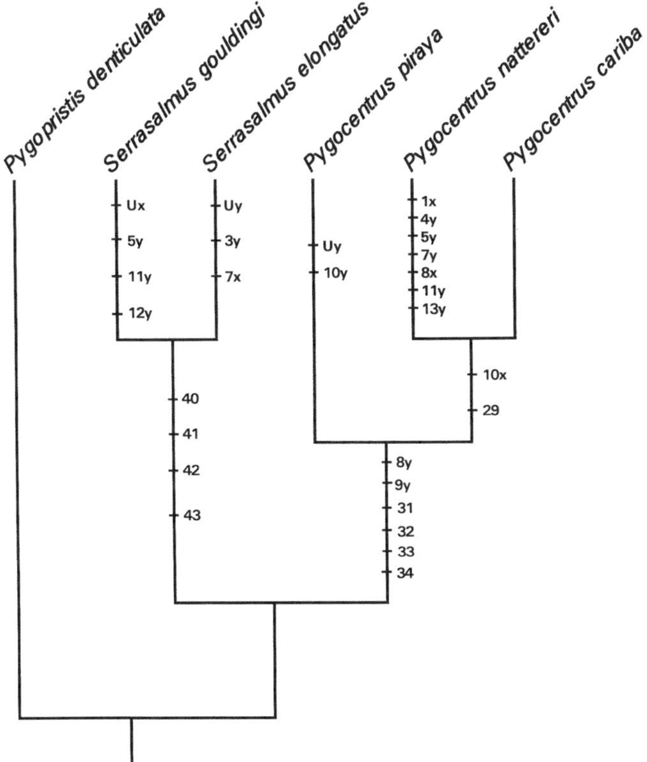

Figure 3.20. Cladogram inferred from partial-warp data and traditional data, showing the distribution of unambiguously optimized characters. The numbers followed by *x* or *y* refer to the numbers assigned to the partial warps, and U denotes the uniform partial warp; *x* and *y* refer to orientation of the partial warps (anteroposterior and dorsoventral, respectively). The numbers 29–43 represent traditional characters.

and Zelditch 1995) and several tests discussed by Emerson and Hastings (1998) do not circumvent systematic judgment. But of course tests circumventing systematic judgment are unavailable for other data as well. Systematic judgment is necessary and cannot be circumvented by rigid rules or algorithms. In our view, the geometric independence of partial warps makes them especially promising for systematic studies because, unlike shape coordinates, partial warps not only take spatial covariances among landmarks into account, they also exhibit less evidence of phylogenetic correlation. Even so, that does not mean that shape coordinates cannot be used in phylogenetic studies. While shape

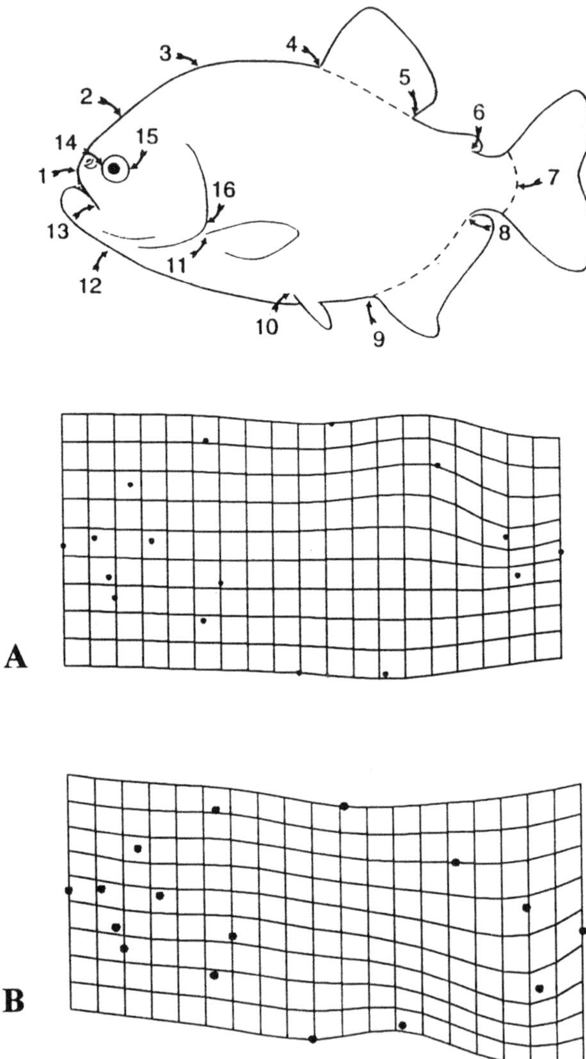

Figure 3.21. Deformations showing ontogenetic trajectories of two features inferred to be correlated in piranhas: **(A)** the deepening of the region between the dorsal and adipose fins relative to the caudal peduncle; **(B)** the deepening of a more ventral posterior region relative to the anterior belly and a change in the ventral profile.

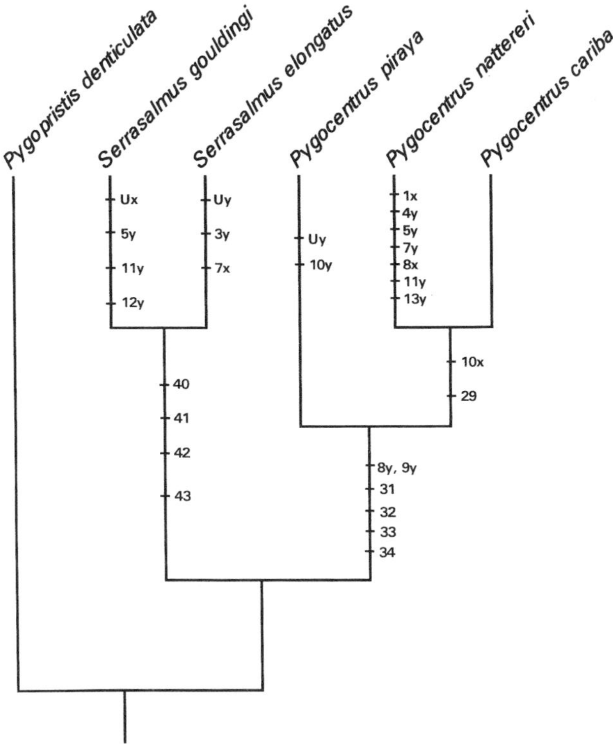

Figure 3.22. Cladogram inferred from partial-warp data and traditional data after the two features inferred to be correlated were recoded as a single character. Symbols showing the distribution of unambiguously optimized characters are as defined for figure 3.20.

coordinates may involve more labor and judgment, they are easily understood so it may not be difficult to judge their independence. And the relatively lesser degree of phylogenetic correlations found in our partial-warp data may be a peculiarity of our piranha ontogenetic data, but even if not, we cannot simply assume that the partial warps are uncorrelated.

CONCLUSIONS

In contrast to several other authors who have considered the matter of quantitative data in systematics, we have come to the conclusion that such data are fully appropriate for systematic, and particularly, phylogenetic studies. Throughout our analysis, we have focused on one issue: the rationale for categorically distinguishing qualitative from quantitative data. We can find no such rationale. Neither the arguments in the literature nor the logic of character analysis suggests any grounds for making that distinction. Instead, we have found objections to specific kinds of variables, which are of limited interest because they are so narrowly focused, or else we have found objections common to all forms of complex data. The latter should be of general interest and should move the debate from the limited domain of morphometrics to phylogenetics at large.

Throughout our analysis, we have approached the subject of morphometric data in the context of the larger domain of phylogenetics. We have asked whether morphometric variables can be analyzed according to the principles generally applied to other data. We conclude that they can be, but we emphasize an important caveat: morphometric variables are not, in and of themselves, equivalent to characters. But even this caveat is not specific to morphometric data and we also argue that qualitative observations are not, in and of themselves, characters. Both morphometrics and qualitative morphology are useful at the starting point of a systematic analysis because both aid in detecting similarities and differences. But the discovery of individual evolutionary novelties (i.e., characters) requires systematic inference. And systematic inferences can be based on *any* observations, whether they are described in qualitative or quantitative form.

The complexity of morphological data means that we must dissect organisms into individual features. This dissection is often difficult and we rarely know, a priori, which lines of dissection correspond to evolutionary units. Yet this dissection is a crucial stage of character analysis. Morphometric data, because they are especially explicit about the features analyzed, and because they force us to be explicit about our lines of dissection, may be a useful general paradigm for complex data. Rather than distinctive or peculiar, morphometric data reveal general characteristics of systematic data.

ACKNOWLEDGMENTS

Our research was supported by National Science Foundation grant DEB-9509195.

REFERENCES

Adams, D. C., and M. S. Rosenberg. 1998. Partial-warps, phylogeny, and ontogeny: A comment on Fink and Zelditch (1995). Systematic Biology 47: 167–172.

Archie, J. W. 1985. Methods for coding variable morphological features for numerical taxonomic analysis. Systematic Zoology 34: 326–345.

Bookstein, F. L. 1989. Principal warps: Thin-plate splines and the decomposition of deformations. IEEE Transactions on Pattern Analysis and Machine Intelligence 11: 567–585.

Bookstein, F. L. 1991. Morphometric Tools for Landmark Data: Geometry and Biology. Cambridge University Press, Cambridge.

Bookstein, F. L. 1994. Can biometrical shape be a homologous character? Pages 198–227 *in* Homology: The Hierarchical Basis of Comparative Biology (ed. B. K. Hall). Academic Press, New York.

Bookstein, F. L., R. E. Strauss, J. M. Humphries, B. Chernoff, R. L. Elder, and G. R. Smith. 1982. A comment upon the use of Fourier methods in systematics. Systematic Zoology 31: 92–95.

Cranston, P. S., and C. J. Humphries. 1988. Cladistics and computers: A chironomid conundrum? Cladistics 4: 72–92.

Creighton, G. K., and R. E. Strauss. 1986. Comparative patterns of growth and development in cricetine rodents and the evolution of ontogeny. Evolution 40: 94–106.

Danser, B. H. 1950. A theory of systematics. Bibliotheca Biotheoretica 4: 117–180.

David, B., and B. Laurin. 1996. Morphometrics and cladistics: Measuring phylogeny in the sea urchin *Echinocardium*. Evolution 50: 348–359.

de Queiroz, K. 1985. The ontogenetic method for determining character polarity and its relevance to phylogenetic systematics. Systematic Zoology 34: 280–299.

Donoghue, M. J., and P. D. Cantino. 1984. The logic and limitations of the outgroup approach to cladistic analysis. Systematic Botany 9: 192–202.

Eigenmann, C. H. 1915. The Serrasalminae and Mylinae. Annals of the Carnegie Museum 9: 225–272.

Emerson, S. B., and P. A. Hastings. 1998. Morphological correlations in evolution: Consequences for phylogenetic analysis. Quarterly Review of Biology 73: 141–162.

Farris, J. S. 1983. The logical basis of phylogenetic analysis. Pages 1–36 *in* Advances in Cladistics, Volume 2 (ed. N. Platnick and V. Funk). Columbia University Press, New York.

Farris, J. S. 1990. Phenetics in camouflage. Cladistics 6: 91–100.

Felsenstein, J. 1981. Evolutionary trees from gene frequencies and quantitative characters: Finding maximum-likelihood estimates. Evolution 35: 1229–1242.

Felsenstein, J. 1988. Phylogenies and quantitative characters. Annual Review of Ecology and Systematics 19: 445–471.

Fink, W. L. 1993. Revision of the piranha genus *Pygocentrus* (Teleostei, Characiformes). Copeia 1993: 665–687.

Fink, W. L., and M. L. Zelditch. 1995. Phylogenetic analysis of ontogenetic shape transformations: A reassessment of the piranha genus *Pygocentrus* (Teleostei). Systematic Biology 44: 343–360.

Garland, T., Jr., and S. C. Adolph. 1994. Why not to do two-species comparative studies: Limitations on inferring adaptation. Physiological Zoology 67: 797–828.

Kluge, A. G., and R. E. Strauss. 1985. Ontogeny and systematics. Annual Review of Ecology and Systematics 16: 247–268.

Lynch, J. M., C. G. Wood, and S. A. Luboga. 1996. Geometric morphometrics in primatology: Craniofacial variation in *Homo sapiens* and *Pan troglodytes*. Folia Primatologica 67: 15–39.

Maddison, W. P. 1991. Squared-change parsimony reconstructions of ancestral states for continuous-valued characters on a phylogenetic tree. Systematic Zoology 40: 304–314.

Maddison, W. P., M. J. Donoghue, and D. R. Maddison. 1984. Outgroup analysis and parsimony. Systematic Zoology 33: 83–103.

McLellan, T., and J. A. Endler. 1998. The relative success of some methods for measuring and describing the shape of complex objects. Systematic Biology 47: 264–281.

Moser, H. G., W. J. Richards, D. M. Cohen, M. P. Fahay, A. W. Kendall, and S. L. Richardson. 1984. Ontogeny and systematics of fishes. American Society of Ichthyologists and Herpetologists, Special Publications 1: 1–760.

Myers, P., B. L. Lundrigan, and P. K. Tucker. 1995. Molecular phylogenetics of oryzomyine rodents: The genus *Oligoryzomys*. Molecular Phylogenetics and Evolution 4: 372–382.

Naylor, G. J. P. 1996. Can partial warp scores be used as cladistic characters? Pages 519–530 *in* Advances in Morphometrics (ed. L. F. Marcus, M. Corti, A. Loy, G. J. P. Naylor, and D. E. Slice). Plenum Press, New York.

Norman, J. R. 1929. The South American characid fishes of the subfamily Serrasalmoninae, with a revision of the genus *Serrasalmus* Lacepède. Proceedings of the Zoological Society of London 55: 781–829.

Patterson, C. 1982. Morphological characters and homology. Pages 21–74 *in* Problems of Phylogenetic Reconstruction (ed. K. A. Joysey and A. E. Friday). Academic Press, London.

Pimentel, R. A., and R. Riggins. 1987. The nature of cladistic data. Cladistics 3: 201–209.

Pogue, M. G., and M. F. Mickevich. 1980. Character definitions and character state delineations: The bête noire of phylogenetic inference. Cladistics 6: 319–364.

Read, D. W., and P. E. Lestrel. 1986. Comments on the use of homologous point measures in systematics: A reply to Bookstein et al. Systematic Zoology 35: 241–253.

Rohlf, F. J. 1997. tpsRegr (computer program), version 1.07. State University of New York, Stony Brook.

Rohlf, F. J. 1998. On applications of geometric morphometrics to studies of ontogeny and phylogeny. Systematic Biology 47: 147–158.

Rohlf, F. J., A. Loy, and M. Corti. 1996. Morphometric analysis of Old World Talpidae (Mammalia: Insectivora) using partial warp scores. Systematic Biology 45: 344–362.

Slice, D. E., F. L. Bookstein, L. F. Marcus, and F. J. Rohlf. 1996. Appendix I: A glossary for geometric morphometrics. Pages 531–551 *in* Advances in Morphometrics (ed. L. F. Marcus, M. Corti, A. Loy, G. J. P. Naylor, and D. E. Slice). Plenum Press, New York.

Stevens, P. 1991. Character states, morphological variation, and phylogenetic analyses: A review. Systematic Botany 16: 553–583.

Swiderski, D. L. 1993. Morphological evolution of the scapula in tree squirrels, chipmunks, and ground squirrels (Sciuridae): An analysis using thin-plate splines. Evolution 47: 1854–1873.

Swiderski, D. L., M. L. Zelditch, and W. L. Fink. 1998. Why morphometrics is not special: Coding quantitative data for phylogenetic analysis. Systematic Biology 47: 508–519.

Swofford, D. L. 1993. PAUP computer program, version 3.1.1. Laboratory of Molecular Systematics, Smithsonian Institution, Washington, D.C.

Thiele, K. 1993. The holy grail of the perfect character: The cladistic treatment of morphometric data. Cladistics 9: 275–304.

Wake, M. 1994. The use of unconventional morphological characters in the analysis of systematic patterns and evolutionary processes. Pages 173–200 *in* Interpreting the Hierarchy of Nature (ed. L. Grande and O. Rieppel). Academic Press, New York.

Wiley, E. O. 1981. Phylogenetics: The Theory and Practice of Phylogenetic Systematics. John Wiley and Sons, New York.

Wilkinson, M. 1995. A comparison of two methods of character construction. Cladistics 11: 297–308.

Zelditch, M. L., and W. L. Fink. 1998. Partial-warps, phylogeny, and ontogeny: A reply to Adams and Rosenberg. Systematic Biology 47: 344–348.

Zelditch, M. L., F. L. Bookstein, and B. L. Lundrigan. 1992. Ontogeny of integrated skull growth in the cotton rat *Sigmodon fulviventer*. Evolution 46: 1164–1180.

Zelditch, M. L., W. L. Fink, and D. L. Swiderski. 1995. Morphometrics, homology, and phylogenetics: Quantified characters as synapomorphies. Systematic Biology 44: 179–189.

Zelditch, M. L., W. L. Fink, and D. L. Swiderski. 1998. On applications of geometric morphometrics to studies of ontogeny and phylogeny: A reply to Rohlf. Systematic Biology 47: 159–167.

Zelditch, M. L., D. L. Swiderski, and W. L. Fink. n.d. Homology, characters, and morphometric data. *In* Fossils, Phylogeny, and Form: An Analytical Approach (ed. J. Adrain and B. Lieberman). Plenum Press, New York. Forthcoming.

PAULA M. MABEE

4

THE USEFULNESS OF ONTOGENY IN INTERPRETING MORPHOLOGICAL CHARACTERS

Interest in the relation between development and evolution has increased dramatically in the past few years (Raff 1996). Great strides are being made in developmental biology, and developmental biologists are increasingly incorporating an evolutionary framework into the interpretation of their results (Holland et al. 1994; Raff 1996; Holland 1997). Questions regarding this relationship, however, are focused differently by those trained in phylogenetic systematics and those trained in developmental biology. Systematists pose questions regarding the usefulness of ontogeny, the observable change in an individual organism through time, in reconstructing phylogeny, such as "What information from ontogeny can be used in reconstructing phylogeny?"—and questions regarding the evolution of character ontogenies, such as "What are the patterns of ontogenetic evolution?" and "How have evolutionary changes in developmental timing affected evolution?" Developmental biologists are also interested in the evolution of ontogenies and have begun to understand the necessity of phylogenetic hypotheses in inferring evolutionary change (Raff 1996). They are often primarily concerned with the role of ontogeny in evolutionary change: "How do ontogenetic mechanisms generate, constrain or channel evolutionary variation?" (Raff 1996).

Although developmental biology is in a renaissance, the field of developmental genetics has not yet reached the stage where the complex anatomical characters of interest to systematists can be explained in terms of known genetic cascades (Holland 1996). The developmental data at hand for systematists consists mainly of transformational data obtained by examining preserved size series of specimens. Moreover, because model organisms are used for most developmental-biology research (Kellogg and Shaffer 1993), systematists hoping to incorporate molecular developmental information into systematic studies are limited to few species. Thus, the usefulness of developmental data for interpreting morphological characters is often limited simply by the lack of developmental data.

Because a temporal dimension is uniquely shared between ontogeny and phylogeny, it has been particularly tempting for biologists to ascribe unusual importance to ontogenetic information. Thus has emerged a set of highly interrelated hypotheses about the relationships between ontogeny and phylogeny. Specifically, ontogenetic transformation has been hypothesized to shed light on phylogenetic character polarity, character state order, and homology—and thereby help systematists in the interpretation of morphological characters. In this chapter, I review these three types of information, all of which are critical in phylogenetic reconstruction and all of which are thought to be available from ontogeny. Throughout I focus on a practical question: What information from ontogeny can be used in reconstructing phylogeny? I also review methods for coding ontogenies as characters in systematics.

The rapid flood of data coming from developmental-biology laboratories on a broad spectrum of model organisms is worthy of attention from systematists. Such information will be valuable in understanding levels of correlation between morphological characters and in formulating character hypotheses. Systematists should be experimenting with different modes of coding and characterizing ontogenetic sequences, so that as our organismal database expands beyond model organisms, developmental data may be used as systematic characters. Congruence tests have the potential to be useful in formulating ontogenetic characters, and this is an unexplored area in developmental morphology and systematics.

ONTOGENY AND CHARACTER STATE POLARITY

Historical Review

The use of ontogeny as a criterion for determining character polarity, distinguishing states that are relatively primitive from those that are relatively derived, has been a matter of active controversy in the past few decades (Lundberg 1973; Nelson 1973, 1978, 1985; Rosen 1982, 1984; Brooks and Wiley 1985; de Queiroz 1985; Kluge 1985, 1988; Kraus 1988; Weston 1988; Mabee 1989b, 1993, 1996; Wheeler 1990; Patterson 1996). Polarity determination is of critical importance to phylogenetic systematists and to all biologists who de-

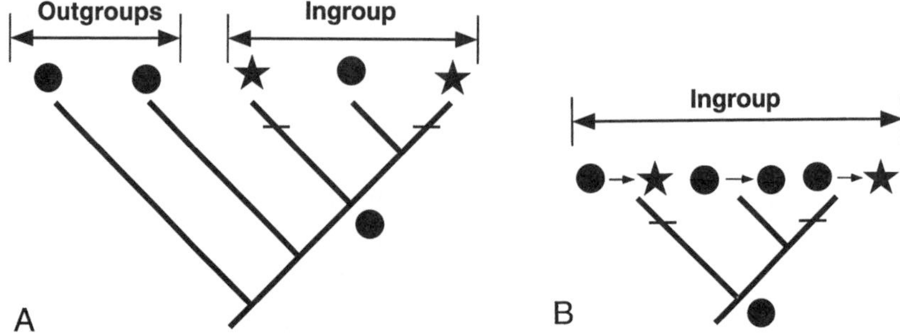

Figure 4.1. Schematic of an application of the outgroup criterion (A) and an application of the ontogenetic criterion (B). The stars and circles represent two different character states. **(A)** By the outgroup criterion, the ingroup is compared with outgroups. Two different character states (star and circle) are present in the ingroup, but only the circle is found in the outgroups. Thus, the ancestral condition for the ingroup is hypothesized to be circle. This diagram represents the star condition as having evolved independently. **(B)** The ontogenetic criterion is employed only in the ingroup. It is observed that the circle precedes the star in development, so the circle is considered more general, the star less general, and the circle is hypothesized as primitive for the ingroup.

pend on phylogenies for comparative studies, because without information on polarity, evolutionary relationships cannot be inferred. This is because only shared derived characters can be used to recognize monophyletic groups in a parsimony framework.

Three methods—outgroup comparison, and to a lesser extent, the paleontological method and the ontogenetic criterion—have been employed in phylogenetic systematics for establishing polarity (Wiley 1981; de Queiroz 1985; Bryant 1991). The outgroup method (figure 4.1A), which is the most widely used in systematics, involves comparison of the character states in the group under study (the ingroup) with homologous character states in closely related groups (Wiley 1981; Maddison et al. 1984). A prior hypothesis of higher level phylogenetic relationships is required to determine the derived conditions (one or more) within the ingroup. The paleontological method involves comparing homologous characters in younger and older fossils. The character state in older fossils in considered to be more primitive than that in the younger fossils. Because stratigraphic order may be violated and because of incomplete sampling, this method is rarely used (Wiley 1981).

The ontogenetic criterion was originally proposed by Nelson (1973, 1978). He stated, "Given an ontogenetic character transformation, from a character observed to be more general to a character observed to be less general, the more general character is primitive and the less general advanced" (Nelson 1978).

The meaning of "generality" is critical to application of the ontogenetic criterion (see "Restatements of the Ontogenetic Criterion," below). The assumptions that underlie the ontogenetic criterion and the theoretical problems of these assumptions have been carefully and explicitly discussed (Lundberg 1973; Brooks and Wiley 1985; de Queiroz 1985; Kluge 1985; Kluge and Strauss 1985). As proposed by Nelson (1973, 1978), the ontogenetic criterion rests on two assumptions: (1) that ancestral ontogenies are retained in descendants (and thus, ancestral characters are always more general), and (2) that derived characters are added terminally. Because Nelson restricted use of the ontogenetic method to cases in which the temporal sequence of ontogenetic transformation is from more general to less general, the latter assumption proves true (figure 4.1B; Brooks and Wiley 1985; Kluge 1985; Kluge and Strauss 1985; Kraus 1988; Weston 1988; Kitching 1992). Although Nelson specifically stated that recapitulatory assumptions involving ontogenetic precedence do not underlie his criterion (Nelson 1973), in fact, as applied to theoretical and actual systematic data (Nelson 1985), terminal addition is assumed (Brooks and Wiley 1985; Kluge and Strauss 1985; Mabee 1993). In practice, the ontogenetic criterion is rarely used to polarize characters (but see Grande 1985; Leis 1986; Brummer et al. 1987; De Pinna 1996). This is not only because the ontogenetic criterion is theoretically and empirically unjustified but also because relevant developmental data are difficult to collect.

Justification of the Ontogenetic Criterion

Many of the proponents of the ontogenetic criterion make claims such as "no biological justification is needed for Nelson's Rule; it is justified by parsimony" (Wheeler 1990; Weston 1994). The parsimony argument that is used to support the ontogenetic criterion (Nelson 1973; Kitching 1992), however, is flawed (Lundberg 1973). Lundberg explained that Nelson argued parsimony at a deeper level than required in a two-taxon approach and thus introduced an additional "step." The difference in the way that parsimony is argued and evolution is viewed is in part a function of differing schools of thought in systematics—namely, phylogenetic systematics versus pattern cladistics. (See Hull 1988 for a description of these schools.) The role of ontogeny in systematics, in fact, is said to be the "kernel of the difference between phylogenetic systematics and pattern cladistics" (Patterson 1994). Unlike in phylogenetic systematics, in pattern cladistics "analysis can proceed without any assumptions about evolutionary process, and without any model" (Patterson 1994). In a similar spirit, Weston (1994) remarked, "It is the observed nested pattern that is important in applying Nelson's Rule, not any developmental or functional model that might be invoked to explain that pattern." Indeed, many proponents of the ontogenetic criterion are fundamentally disinterested in a biological or evolutionary justification. Parsimony arguments, though proven flawed, are considered adequate justification.

Some of the proponents of the ontogenetic criterion claim that ontogeny is a "direct method" or an "independent criterion" for establishing character polarity, in that the ancestral condition can be recognized by comparing the ontogenies of two putative sister taxa (Nelson 1973, 1978; Nelson and Platnick 1981; Rosen 1982; Weston 1988). It is direct in the sense that an investigator can simply watch the development of living taxa and see features transform from more to less general (i.e., from less to more derived). In contrast, use of the outgroup method involves at least a three-taxon comparison and a prior hypothesis of higher level phylogenetic relationships. Because an investigator relies on inference, the outgroup method is in that sense indirect. Oddly, there is a dispute among the proponents of the ontogenetic method regarding whether it is actually direct or indirect. Wheeler (1990), for example, does not consider Nelson's rule to be a direct method because it relies on the choice of a second taxon for comparison—a point with which Weston (1994) disagrees. Although outgroup and ontogenetic methods are seen not as opposing but as complementary methods (Wheeler 1990; Weston 1994; Meier 1995), they are based on very different assumptions.

De Queiroz (1985), Bryant (1991), I (Mabee 1993), and others have argued that the ontogenetic method lacks a theoretical basis. Specifically, there is no biological reason to expect that ontogenetic generality, however defined, will reflect the direction of evolutionary change. This is in contrast to the outgroup criterion, which is expected to work when descent occurs with modification and modifications are retained in descendants. On the one hand, proponents indicate that a biological justification is not needed (Weston 1994), yet on the other they indicate biological reasons for why they expect the ontogenetic criterion to work: "There is a good biological explanation for the overwhelming parallel between direction of ontogenetic transformation and phylogenetic polarity as indicated by the relative generality of characters. That is that characters linked by ontogenetic transformation are epigenetically constrained" (Weston 1994). No empirical data have been provided to support this statement, and given the general paucity of data on developmental mechanisms constraining morphological characters, this is not surprising. It is likely, however, that some ontogenetic characters are constrained and others are not.

Restatements of the Ontogenetic Criterion

To apply the ontogenetic criterion, greater generality must be sorted from lesser generality. The lack of an explicit definition of "generality" by Nelson has led to restatements of the ontogenetic criterion by many: de Queiroz (1985), Weston (1988, 1994), Wheeler (1990), Williams et al. (1990), Bryant (1991), Kitching (1992), De Pinna (1994), and Patterson (1996). This subject was partially reviewed in Meier (1997). For example, in contrast to Weston (1988, 1994), who expressed that the only relevant information is the relative generality of the character states, De Pinna (1994) presented directionality, the "order in which character states follow one another," as the most relevant ontogenetic information. Bryant (1991) pointed out that retention of ancestral ontogenetic character states in the ontogeny of descendants is the crucial and testable auxiliary assumption that is made in his own restatement of the ontogenetic method. These elaborations, restatements, and fine-tuning of Nelson's method have led to less and less uniformity in application of ontogeny to polarity determination. Therefore, it has become increasingly difficult to make empirical comparisons between the ontogenetic criterion (or criteria) and the outgroup criterion. Meier (1997) recoded data sets from an earlier study of his (Meier 1995) and from three other studies (Kraus 1988; Wheeler 1990; Mabee 1993), according to what he viewed in the literature as "four different interpretations" of "generality" in the ontogenetic criterion. It is unclear from his study, however, which interpretation performed the best according to specified criteria, though he did find that the "biogenetic rule" interpretation yielded ancestors that require somewhat more homoplasy than do others. Meier (1997) argued that the ontogenetic criterion deserves further empirical consideration because it (some unspecified version) "found the same tree roots as did outgroup comparison" for three out of four empirical data sets that he reanalyzed.

Meier, however, does not indicate the biological reasons that lead one to expect that ontogeny would predict evolutionary polarity. Regardless of how "generality" is defined, on what grounds would developmental change be expected to reflect the direction of evolutionary change? Meier (1997), in fact, disagrees with the parsimony argument that many use to justify the ontogenetic criterion. However interesting the results of empirical tests, they are secondary to the deeper evolutionary arguments and assumptions underlying polarity criteria.

Empirical Evaluations of the Ontogenetic Criterion

Prior to the empirical study in which Mabee (1989b, 1993, 1996) examined the performance of the ontogenetic criterion on centrarchid fishes, the ontogenetic criterion had been evaluated from a small set of morphometric data on fossil bivalves (Miyazaki and Mickevich 1982) and from a larger, more diverse set of character data on salamanders (Kraus 1988). Kraus (1988) compared topological similarity, level of resolution, and degree of ambiguity of trees polarized on the basis of ontogeny and those polarized by means of outgroups and found that the ontogenetic criterion was as successful as the outgroup criterion. Wheeler (1990) and Meier (1997), both using insect data, found empirical support for the ontogenetic criterion. Meier (1997) used a data set of 3 egg, 58 larval, and 27 adult characters of Sepsidae (Diptera) and a method of mapping character change on "ontogenetic" and "reverse ontogeny" trees (from Mabee 1993). He found that terminal addition accounts for 96% of the ontogenetic change to which the ontogenetic criterion was applicable (see table 5 in Meier 1997). This suggests that the ontogenetic criterion is useful for polarizing Meier's sepsid data set.

As Meier (1997) pointed out, empirical tests differ significantly in their interpretation of "generality" in Nelson's (1973) criterion, and he usefully standardized four different empirical data sets in his review. He found that the outgroup criterion is applicable to about as many characters as the ontogenetic criterion for several of the data sets reviewed. Meier (1997) also discovered that the ontogenetic criterion identified the same roots as the outgroup criterion. Although he interprets this finding as support for the notion that the ontogenetic criterion has a sound theoretical foundation, he does not suggest what that foundation might be, nor does he demonstrate what the expectation would be due to chance alone. In another study (one not reviewed in Meier 1997), Hibbett et al. (1993) tested the ontogenetic criterion on morphological characters from fungal sporocarp ontogeny, and because of some nonterminal additions, deletions, and de-differentiation in development, the results did not support the use of ontogeny as a criterion for assessing evolutionary polarity.

To test the ontogenetic criterion, ideally one would use for comparison the single most parsimonious phylogeny that is based on data completely independent of ontogeny (Mabee 1989b, 1993). Alternatively, in the absence of such a tree, all possible phylogenetic hypotheses might be examined. In a study of centrarchid fishes (Mabee 1993), I circumvented the lack of a single most parsimonious tree, as well as the logistic constraints of character state interpretation for all possible trees (for centrarchids, 8.7×10^{36} rooted trees are possible), using the following approach. First, I did a phylogenetic analysis based on 56 morphological characters, excluding all ontogenetic information, and I used the outgroup criterion to root the tree. This analysis resulted in more than 100 equally parsimonious trees, and one that was consistent with other estimates of centrarchid relationships was chosen as a working hypothesis. On this tree, the ontogenetic criterion was "correct" in its polarity for about one-third (32%) of the total phylogenetic change. In other words, the direction of ontogenetic change and evolutionary polarity are congruent for about one-third of the phylogenetic change (see terminal addition, figure 4.2). Given the number of other equally parsimonious trees for centrarchids, however, as well as the enormous

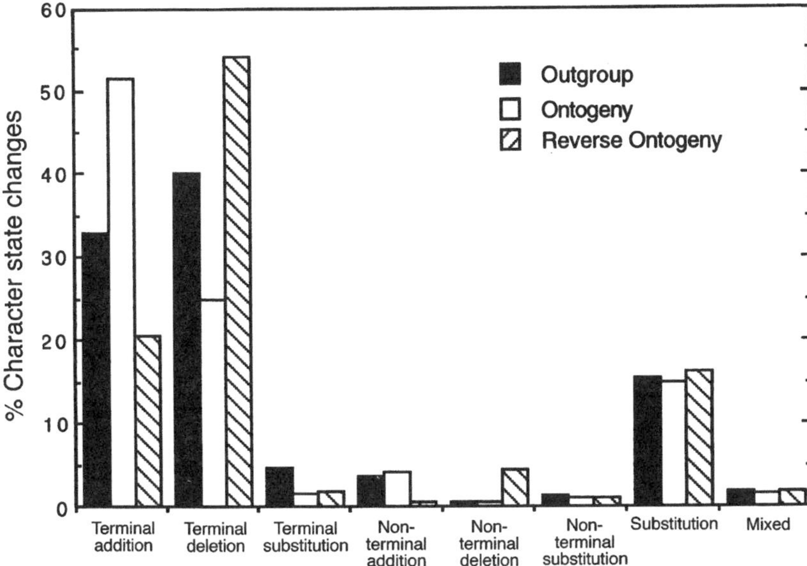

Figure 4.2. Frequencies with which different classes of evolutionary changes are observed in ontogeny. "Mixed" refers to changes that could be interpreted in more than one class. Adapted from Mabee (1993).

number of total possible trees, it seemed likely that there might be a tree or group of trees for which the ontogenetic criterion would correctly polarize much more of the character change. If the ontogenetic criterion were evaluated on the single true tree representing centrarchid evolution, would it predict polarity correctly for 100% of the character state change?

The question then became how to discover the single tree out of the 8.7×10^{36} possible that is most consistent with the ontogenetic criterion, the tree for which the maximum amount of character state change is correctly interpreted by the ontogenetic criterion. I constructed a "boundary condition" tree by polarizing all characters on the basis of the ontogenetic criterion and ordering multistate characters according to the order in which they appeared during ontogeny. Interpreted in light of this tree (one that is exceedingly improbable in light of probable outgroups), the ontogenetic criterion correctly polarizes about half (52%) of the character state change (figure 4.2). I also calculated the tree that implied the minimum amount of character state change that could be correctly polarized by the ontogenetic criterion and found a lower limit of 21% (figure 4.2). My test of the ontogenetic criterion was independent of phylogeny in the sense that, for *any* of the 8.7×10^{36} possible trees for centrarchids, the ontogenetic criterion can predict the direction of evolution for, at most, only 52% of the character change. I concluded that the ontogenetic criterion is not an accurate predictor of evolutionary direction (Mabee 1989a, 1993). It is not empirically justified as a reliable or general indicator of phylogenetic polarity.

When the ontogeny of a paedomorphic character (terminal deletion, figure 4.2) is compared with the ontogeny of a character in which the ancestral adult state has been retained, the ontogenetic criterion will lead to the incorrect recognition of a derived paedomorphic condition as primitive (Lundberg 1973; Wiley 1981). Instances of paedomorphosis in both plants and animals have been well documented (e.g., Hanken 1984; Mishler 1988). The ontogenetic criterion assumes that paedomorphosis and all other evolutionary modifications except terminal additions are infrequent. In addition to giving the wrong polarity in cases of terminal deletion or paedomorphosis, the ontogenetic criterion is typically inapplicable or ambiguous in cases where the ontogenetic transformation does not progress from greater to lesser generality—as in cases of terminal substitution, nonterminal deletion, nonterminal substitution, nonterminal addition, and substitution. Although it might be argued that the ontogenetic criterion cannot be tested in sequences that do not proceed from more to less general (in cases termed "substitutions"—where homologies among ontogenies cannot be ascertained), these ontogenies still represent a class of evolutionary changes for which polarity cannot be assessed from ontogeny. The ontogenetic criterion is applicable only in cases where its assumptions hold true—that is, in cases of terminal addition. Considering the maximum relative frequency of that class of character state change (see figure 4.2), ontogeny is not a reliable criterion for establishing phylogenetic polarity.

Criticisms of my studies (i.e., Mabee 1989a, 1993) have centered around a misunderstanding of comparison methods and interpretation of "generality." Kitching (1992), for example, misunderstood the comparisons with outgroup trees. He stated, "[H]er argument is flawed because she assumes that the trees found by outgroup comparison are the 'correct' trees" (Kitching 1992). This reflects Kitching's misunderstanding of the methods that I used to evaluate the ontogenetic criterion, by which I essentially compared the utility of the ontogenetic criterion for *all* possible centrarchid trees. Patterson (1996) called my analysis into question on the basis of his reinterpretation of the ontogenetic criterion, specifically my designation of "general" states in centrarchid characters. The intent in these analyses was to conform as closely as possible to the original formulation of the ontogenetic criterion as stated by Nelson (1978), and I therefore discussed the polarity of many of the characters with Nelson. Patterson (1996) did not accuse me of misinterpreting Nelson, nor did he claim that his interpretation of the ontogenetic criterion is closer to Nelson's. Moreover,

Patterson did not explain how his interpretation of the ontogenetic criterion differs from mine (or Nelson's); only by examining Patterson's application of the ontogenetic criterion to characters do the fundamental differences in our interpretations become apparent. Most basically, Patterson and I designate different ontogenetic levels as "general": to Patterson (1996), "general" meant the earlier ontogenetic absence of characters, whereas I considered it to mean the earliest ontogenetic character state that is also observed as a terminal ontogenetic state in the ingroup. Only my application demands that the states considered for polarization be relevant to the phylogenetic question at hand: centrarchid interrelationships. Again, I find that differing interpretations of what Nelson meant by "generality" serve only to cloud the critical issue: the lack of a theoretical basis for the ontogenetic criterion.

Even though the ontogenetic method has been demonstrated to be deeply flawed both theoretically and empirically, the belief that ontogenetic transformation represents a uniquely preserved record of phylogenetic transformation is deeply rooted. For example, De Pinna (1994) stated, "Although the parallelism between ontogeny and phylogeny is yet to be tested according to precise cladistic methods, I believe that within the massive assemblage of developmental facts amassed during the last 200 years, there is a strong sign that such a parallelism actually exists." Apparently this belief, that ontogeny contains special phylogenetic information (other than simply character information), will continue to reappear in the biological literature (Gould 1977).

ONTOGENY AND CHARACTER STATE ORDER

Developmental data are a source of information from which character state relationships are sometimes inferred in systematic studies (Hauser and Presch 1991). Although ontogenetic data are often used implicitly (or rarely, explicitly) to order character states, the two components of transformational data, order and polarity, are often confused. Polarity and order often are not distinguished in systematic studies, even though they are separate aspects of character state hypotheses (Swofford et al. 1996). The terms "ordered" and "unordered" apply specifically to characters with more than two states (multistate characters). These terms refer to hypotheses of evolutionary relationship among character states; they do not imply evolutionary direction. An ordered character state hypothesis is analogous to an unrooted taxon tree (Lundberg 1973), because topographic relationships among states are inferred but polarity is not specified.

In an initial hypothesis of homology among character states, all states are assumed to be homologous, but no particular relationships are specified. That is,

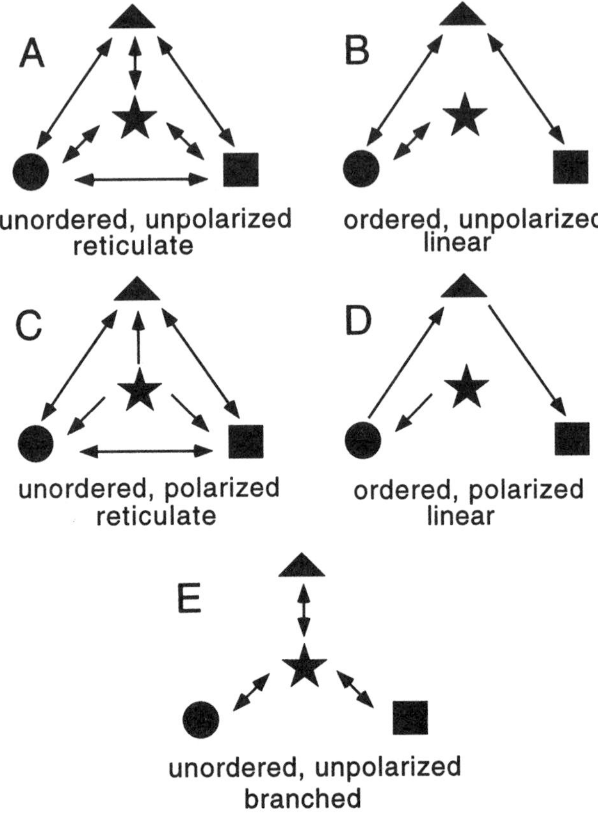

Figure 4.3. Hypothesized relationships among four character states (star, circle, triangle, and square). These states are considered homologs, and thus states of a single character. The different relationships of order and polarity that can be specified among homologous states are diagrammed.

neither character state order nor direction is implied; the character states are unordered and unpolarized (figure 4.3A). An unordered but polarized character (figure 4.3C) has a hypothesized primitive state (the star) from which another unspecified state was derived. In an ordered, unpolarized character (figure 4.3B), some relationships among character states are hypothesized. These hypotheses are based on an "outside" source of information such as development, or on particular assumptions about evolutionary processes. Finally, an ordered, polarized character (figure 4.3D) contains information both about evolutionary direction (polarity) and character state relationships (order). Ordered character states may take one of three configurations: linear (figure 4.3B, D), branched (figure 4.3E), or reticulate (figure 4.3A, C).

Hypotheses of polarity and order restrict or change the number and topology of resultant phylogenetic trees. Specification of character states as ordered or unordered involves restrictive and different assumptions about evolutionary processes (Miyazaki and Mickevich 1982; O'Grady and Deets 1987; Mabee 1989a; O'Grady et al. 1989; Hauser and Presch 1991). Generally, ordered character states restrict tree numbers and topologies more than unordered characters do (Mabee 1989a; but see Hauser and Presch 1991), and it is critical that investigators are careful and explicit about the criterion used to base the ordering upon. Developmental data is generally used only implicitly to order character states. Although there is a tendency to seek phylogenetic information in the transformations observed in development, there is no reason to expect a correspondence between the two. Empirically it appears that sometimes there is a parallel between observed ontogenetic change and observed phylogenetic change (Meier 1997). In other cases, the order implied by ontogeny is incongruent with the order of character state transformation implied by phylogeny (Mabee 1989a, 1993). Therefore, the use of ontogeny to order characters must be explicitly justified. Further studies on the parallel or lack thereof between observed character state change and character change as inferred from phylogenetic trees will shed more light on this issue.

ONTOGENIES AS CHARACTERS

Beyond the problematic use of ontogeny in polarizing or ordering characters, the question of how character ontogenies themselves can be used in phylogenetic analyses has remained unanswered. The ontogeny of morphological characters typically refers to the observed sequence of morphological change or transformation. Whether ontogenies are to be used as characters in phylogenetic studies by systematists or are to be compared in context of a phylogenetic tree by developmental biologists, they must be homologous. That is, they must be similar due to common ancestry. Ontogenetic characters are individuated parts of the ontogeny of the whole organism that have evolved with respect to their homologs in another taxon. In cases where ontogenetic character recognition is straightforward (confounding factors are discussed under "Coding Multiple Character Ontogenies [Event Sequences] for Phylogenetic Inference"), how should they be coded?

Coding a Character Ontogeny for Phylogenetic Inference

The difficulty of using ontogenies directly in phylogenetic analyses is rarely appreciated. In most studies, when ontogenies are coded as characters, either they are reduced to simple presence of transformation *(a → b)* or absence of transformation *(a → a)*, or they are given different character codes to represent different patterns of transformation (e.g., *a* → *b* coded as "0"; *a* → *b* → *c* coded as "1"; *b* → *c* coded as "2", etc.) (Wake and Larson 1987; Mabee 1989a, 1993; Tyler et al. 1989; Wake 1989). Other so-called ontogenetic characters are taken from a single early life-history stage of an organism (semaphorant) and are therefore static characters, incorporating no transformational data (Meier 1995). The danger associated with both these methods of coding is that homology information among character states may be lost.

The approaches and problems involved in coding ontogenies as characters are directly analogous to those involved in coding allozymes (Mabee and Humphries 1993). Just as alleles may be coded as separate characters with presence or absence as the states, parts of ontogenetic sequences may be coded as separate characters with presence or absence as the states (Klompen and O'Connor 1989). Consider the example in which five different ontogenetic states *(r–v)* have been observed in a character transformation for seven taxa (A–G) (table 4.1). If each of these five states is considered a separate character, with presence (1) and absence (0) as the possible states (table 4.2), phylogenetic analysis results in one most parsimonious tree (figure 4.4). The common hypothetical ancestor of taxa A–E, however, lacks an ontogeny! This implies that either the ancestor had an ontogeny that is not found in any extant taxon or that the character was not appropriately coded. Given that conditions in extant taxa are inherited from homologous ones in their ancestors, we are forced to conclude that character conception is the problem. The ontogenetic characters

Table 4.1

Observed ontogenetic character transformations

Taxon	Ontogenetic character transformation
A	*t* → *u* → *v*
B	*t* → *u* → *v*
C	*t* → *u*
D	*t*
E	*r* → *v*
F	*s*
G	*s*

Table 4.2

Data matrix for observations in table 4.1, where each possible condition in an ontogenetic transformation is considered to be a character

Taxon	Ontogenetic character				
	r	*s*	*t*	*u*	*v*
A	0	0	1	1	1
B	0	0	1	1	1
C	0	0	1	1	0
D	0	0	1	0	0
E	1	0	0	0	0
F	0	1	0	0	0
G	0	1	0	0	0

Character states: present (1); absent (0).

conceived here are not independent, but dependent, and they violate the assumptions of phylogenetic analysis.

Another approach to coding character ontogenies involves coding different ontogenies as different states. Just as an allozyme locus may be coded as the character and its allelic combinations as the states, so a character ontogeny may be coded as the character and the sequence combinations as the states. If this approach were used to recode the ontogenies in table 4.1, each unique combination of the five different states observed in the character ontogeny would be

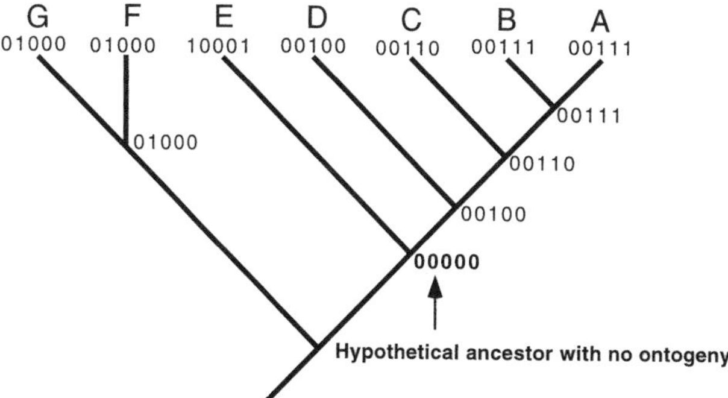

Figure 4.4. Phylogenetic tree of taxa A–G resulting from an analysis in which each possible condition in an ontogenetic transformation is considered to be a character. Character states: present (1); absent (0). Data are from table 4.2.

Table 4.3

Codes for observations in table 4.1 if the ontogenetic transformation as a whole is considered the character and each unique combination of ontogenetic states is considered a separate state

Taxon	Ontogenetic character states	Code
A	$t \rightarrow u \rightarrow v$	0
B	$t \rightarrow u \rightarrow v$	0
C	$t \rightarrow u$	1
D	t	2
E	$r \rightarrow v$	3
F	s	4
G	s	4

considered a different character state and coded accordingly (table 4.3). If this matrix is analyzed with the character unordered, 77 equally parsimonious trees are found (Mabee and Humphries 1993). Such character states may be ordered according to the investigator's assumptions about the evolutionary process (Mabee 1989a; Hauser and Presch 1991). The value of this approach is that the biological unity of the ontogenetic character is recognized. The problem is that information about shared homology is lost—e.g., although different taxa share character states, this information is lost because different state combinations are given different codes.

Mabee and Humphries (1993) presented a method, involving the step matrix option in PAUP, for maximizing the retention of character homology while retaining a biologically sound definition of characters. This allows use of the entire ontogenetic sequence as the character in contrast to coding the condition at each stage (semaphorant) as a separate character. The benefit of coding

Table 4.4

Character step matrix of observed ontogenetic character state combinations with pairwise distances

	tuv	*tu*	*t*	*rv*	*s*
tuv	—	1	2	3	4
tu	1	—	1	4	3
t	2	1	—	3	2
rv	3	4	3	—	3
s	4	3	2	3	—

See also figure 4.5.

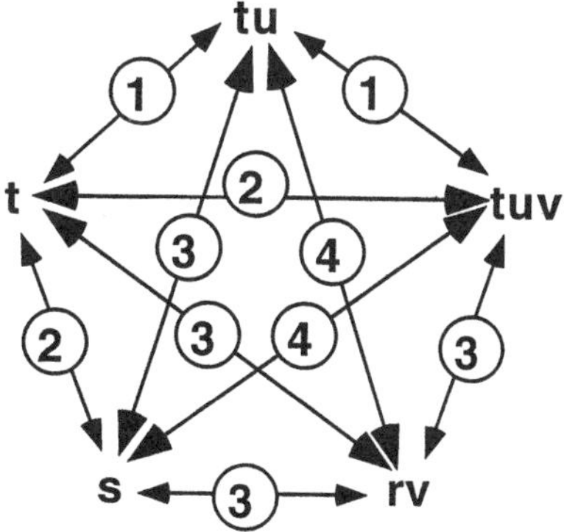

Figure 4.5. Reticulate character state tree representing the relationships between different ontogenetic character transformations in table 4.1. Numbers in circles are the number of steps (see table 4.4). Redrawn from Mabee and Humphries (1993).

character ontogenies into a step matrix is that combinations of observed conditions within ontogenies may be coded as character states. A step matrix for the ontogenetic character transformations observed in table 4.1 is shown in table 4.4. A step matrix is essentially a character state by character state distance matrix in which the distance hypothesized between different states reflects an investigator's hypothesis of order, the likelihood of evolutionary transformation. With a step matrix, hypotheses of character state homology that imply reticulate character state trees can be encoded, as can linear or branching character state trees. Ordered linear or branched character state hypotheses cannot reflect all shared alleles among allelic combinations, only a reticulate character state tree will reflect these shared homologies (figure 4.4). The distances on this character state tree are calculated assuming equal weighting of loses and gains. For example, a distance of 1 is hypothesized between $t \rightarrow u$ and $t \rightarrow u \rightarrow v$. This means that if $t \rightarrow u$ is the primitive condition, v is gained in one step, but if $t \rightarrow u \rightarrow v$ is the primitive condition, v is lost in one step. Other scaling or weighting assumptions may be employed to reflect frequency information (advocated by Wiens 1995) in the case of polymorphic ontogenies (species with more than one transformation pattern for a particular character).

Reticulate character state trees (figure 4.5) can be translated into a step matrix (table 4.4) and input into the data matrix (see part c of the appendix in Mabee and Humphries 1993). The unity of the character ontogeny is preserved (unlike in the ontogenetic-state-as-character approach), as is the homology information of the shared ontogenetic conditions (unlike in the whole-ontogeny-as-character approach). Homology among ontogenetic sequences is retained and the ontogenetic sequence as a whole can be considered the character. Because ontogenetic sequences need not be broken into their component parts for phylogenetic analysis, the problems incurred in assuming that the components are independent are avoided. That is, reconstructing ancestors lacking ontogenies is avoided with this step matrix approach.

Coding Multiple-Character Ontogenies (Event Sequences) for Phylogenetic Inference

There are many cases where recognition of ontogenetic characters is not straightforward. Specifically, there may be temporal sequences of morphological events that may or may not be causally or mechanistically connected. The problem essentially involves delimiting the character ontogeny from the ontogeny of the organism as a whole. Given that organisms are highly integrated collections of ontogenies, but that characters to be used in phylogenetic analysis should be independent of one another, the practical difficulty of separating out relatively independent ontogenetic characters presents itself. What criteria can be used to delimit ontogenetic characters?

The difficulty of recognizing ontogenetically integrated units, or characters, can be illustrated with data from skull ontogeny in fishes. The developmental timing of cranial ossification of *Danio rerio* has been described in detail by Cubbage and Mabee (1996). The sequence of formation of every bone in the skull can be diagrammed linearly (figure 4.6). Thus, ceratobranchial 5 (a gill arch bone) ossifies first, followed by the opercle (gill cover bone), the parasphenoid (roof of mouth), the branchiostegal rays (gill membrane supports), and so on. Looking at these data for potential systematic characters, we might ask, "Is every bone in the skull a separate character or might regional groups of bones, for example, be appropriate characters?" The opercular region of the skull is often considered to include the opercle, subopercle, interopercle, and preopercle. The interopercle and subopercle form in close succession and are fairly widely separated in time of formation (as measured by age and size) from the other two opercular bones. Is the appropriate character the "opercular series," which includes all four bones in their sequence of formation, or the "interopercle–subopercle" sequence? That is, is the temporal sequence itself a heritable feature that has homologies in other taxa? Or most simply, is sequence unimportant; and the presence or absence of individual bones the characters? If

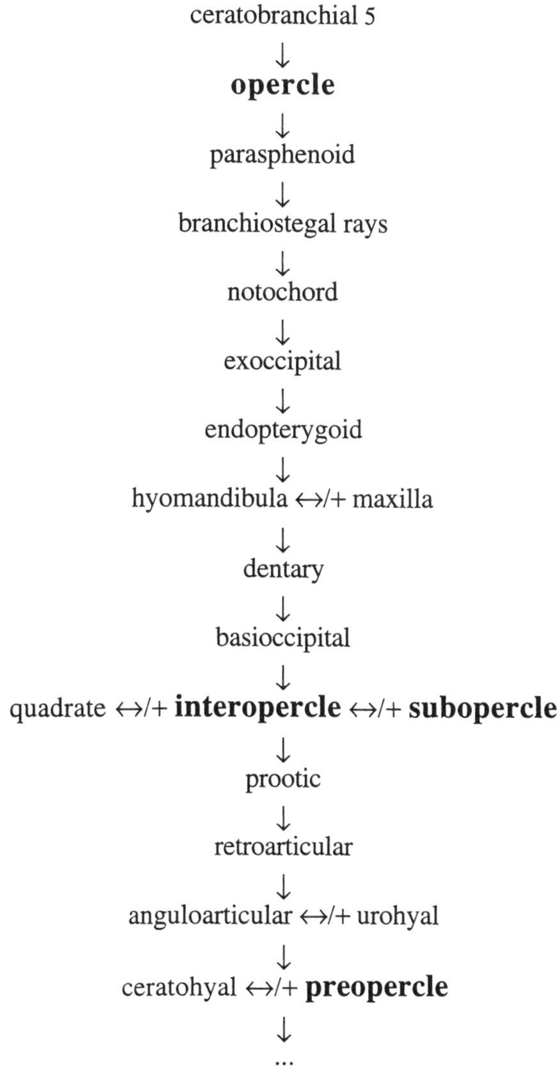

Figure 4.6. Part of the most common sequence of cranial ossification in *Danio rerio*. Arrows reflect the progression of ossification. Terms in bold are bones of the opercular series. Either of the elements connected by "↔/+" may develop before the other, or they may develop simultaneously. For example, after the endopterygoid develops, either the hyomandibula and then the maxilla develops or vice versa, or these two bones develop at the same time. Adapted from Cubbage and Mabee (1996).

they are induced by the same developmental mechanism, individual bones would not be independent characters, a requirement for phylogenetic analysis.

An ontogenetic sequence coding method for multiple characters, termed temporal "event sequences," was described and used by Velhagen (1995, 1997) and has been used by Mabee and Trendler (1996) and Smith (1996). This method must be distinguished from methods for coding individual characters, such as step matrixes (described above).

Velhagen (1997) distinguished the "transformation sequence," the change of a structure from one form into another (e.g., opercle smooth → opercle serrate) from the "event sequence," the ordinal/temporal succession of events (e.g., opercle ossifies → preopercle ossifies → subopercle ossifies). According to Velhagen's method, event sequences are broken into series of pairwise sequences: the order of two events relative to each other. Thus, the opercle–preopercle sequence, the preopercle–subopercle sequence, and the opercle–subopercle sequence would each be characters. Three states are possible for each character: first event occurs first (coded as "1"), second event occurs first ("2"), or both events occur simultaneously ("?"). Thus, if the opercle ossifies before the preopercle, the character state would be coded as "1"; if the preopercle ossifies before the opercle, it would be coded as "2"; and if the two areas ossify simultaneously, the code "?" would be used. In such manner, every temporal series of events can be broken into pairwise segments, coded, and compared with homologous event sequences in phylogenetic analysis. These characters are weakly concordant with other phylogenetic hypotheses in Velhagen's (1997) and Smith's (1996) analyses, and thus are not necessarily "good" characters. The drawback in using Velhagen's method (pairwise event sequences as characters) is that it assumes that the characters are independent of each other. They are not; as Velhagen (1997) pointed out, "to various degrees, this assumption is not true logically or biologically." This approach of atomizing sequences is thus problematic as a way to generate independent characters for phylogenetic analysis. Still, it is a useful exploratory approach, as it may allow an investigator to determine what parts of a larger sequence are conserved or changing (Mabee and Trendler 1996) and which components may be integrated. Further developmental information will elucidate the degree of integration among different morphological structures. This, in turn, will allow investigators to appropriately recognize and code characters.

Given that characters are inferences of homology and not simple observations, the criteria that are typically used to infer homology should be examined. Morphological homology is operationally recognized in many phylogenetic studies by employing traditional criteria: similarity of structure, position relative to other structures, and development (Wiley 1981). In the study of development itself, however, it is difficult to use these. First, many features appear superficially similar in structure due to their small size or lack of differentiation. Second, assessing positional similarity is difficult because of the lack of development (absence) of reference or "landmark" structures. Third, formulating hypotheses of homology of ontogenetic characters by using ontogeny to hypothesize homology leads to a potential misinterpretation of the way in which ontogenies have evolved (Mabee 1993). Given the shortcomings of standard operational criteria, what criteria *can* be used to establish homologies of developmental sequences? Developmental sequences may also be delimited on the basis of their temporal and spatial coherence (illustrated by the opercular series, above) (Mabee 1993) or integration, if we recognize that developmental integration itself may change through evolutionary time (Fink and Zelditch 1996). But how do we recognize these integrated "natural" units (Wagner 1996)? Analogous to constructing phylogenetic hypotheses, ontogenetic character hypotheses may be evaluated (1) with information on underlying developmental processes and (2) by congruence with other types of character data within the context of a phylogenetic analysis.

The importance of understanding underlying developmental processes in the selection and interpretation of morphological developmental sequences was first emphasized by Alberch (1985) and repeated by others (Raff and Wray 1989; Wray and McClay 1989; Hall 1990). Alberch (1985) stated that only "causal" sequences (as opposed to temporal sequences) are useful in systematic and heterochronic analyses. He argued that only causal sequences (e.g., in the character ontogeny $a \rightarrow b \rightarrow c$, that a induces b and that b induces c) are conserved in phylogeny and thus are the only developmental sequences of value to systematists. In essence, he maintained that a character should be developmentally integrated, not composed of two or more different characters. As Alberch (1985) pointed out, however, identifying causal sequences is difficult for any phylogenetic study involving the usual broad spectrum of complex morphological characters, because so little mechanistic information is available. As additional developmental data become available, this will be a more tractable option.

Congruence studies for the purpose of evaluating the validity and utility of ontogenetic characters in phylogenetic studies have not been done. Congruence studies seek out common phylogenetic patterns in multiple, independent data sets (Hillis 1995; Miyamoto and Fitch 1995). Studies in which larval and adult data sets of metamorphic taxa are compared have been done (Judd 1998), but there is a need for studies in which different hypotheses of characters are tested. The formulation of ontogenetic characters can potentially be tested using the congruence approach, and this is an important goal for future ontogeny/phylogeny studies in systematics.

Criticisms of Ontogenies as Characters

Weston (1988) and Kitching (1992) criticized de Queiroz's proposition that ontogenetic transformations are characters (de Queiroz 1985), on the grounds that his definition of characters was ambiguous. De Queiroz (1985) asserted that ontogenetic transformations, which he defined as parts of characters, are the characters of choice in phylogenetic analysis. He contrasted these with "instantaneous morphologies," which are parts of a character ontogeny viewed at a single point in a life cycle. He contended that if one accepts ontogenetic transformations as characters, there can be no ontogenetic method for polarity determination; the ontogenetic transformations within a character say nothing about the evolutionary polarity of that character. One of Weston's (1988) objections was that, taken to the extreme, de Queiroz's approach would view an entire organism as a single character. Kitching's (1992) objections were unclear but focused generally on ambiguity and confusion in de Queiroz's definition.

Similarly, Patterson (1994) criticized my definition of an ontogenetic character (Mabee 1993), which corresponds closely with that of de Queiroz (1985):

> Characters and character states are hypotheses of homology. They are individuated (delimited) parts of the ontogeny of the whole organism that have evolved with respect to their homologues in another taxon. "Character" and "character state" are relative terms and represent units of phylogenetic information (hypotheses of homology) at different levels; both may represent the evolving parts of organism ontogeny. (Mabee 1993: 190)

Patterson stated that this evolutionary interpretation of characters as process leads to "paradox and perplexity"; he went on to say:

> If we are willing to discriminate pattern from process and to treat cladistic analysis in morphology as pattern analysis, then I remain convinced that the ontogenetic argument, generalized as the observation that absence is more general than presence, is enough to polarize some proportion of the characters in any matrix. It follows that cladistic (phylogenetic) analysis can proceed without any assumptions about evolutionary process, and without any model. (Patterson 1994)

Patterson conflated issues surrounding the ontogenetic criterion for polarizing characters (reviewed above) with those involving ontogenies as characters in phylogenetic analyses. He was unwilling to allow "process" information to enter into character formulation. Given the renaissance in developmental biology and the great number of new characters potentially available to phylogenetics, it seems premature to exclude ontogenetic data.

ONTOGENY AND HOMOLOGY

Similarity of development has long been considered an important, if not the most important, criterion for inferring homology. Some examples of this are well known: homology between the ear ossicles of mammals and the jaw bones of fishes was clarified from similarity in development, as was the homology between flower petals and leaves. Congruence between hypotheses of homology of morphological structures (vertebrae) derived from positional information and homology as inferred from *Hox* gene expression (positional information at a molecular level) (Burke et al. 1995) gives confidence that development is a reliable predictor. On the other hand, it is well recognized that development is not a foolproof criterion for inferring homology (Wagner 1989a, 1989b; Raff 1996), and developmental data are often paradoxical to interpret because morphological characters may be dissociated from their underlying developmental mechanisms (for review, see Müller and Wagner 1996). For example, homologous genes are involved in the initiation of eye morphogenesis in vertebrate and invertebrate eyes (Quiring et al. 1994; Halder et al. 1995). It is difficult to reconcile this similarity with the enormous disparity in eye structure. Apparently eye morphogenesis has remained under conserved genetic control over vast periods of evolutionary time, while the actual morphology of eyes has evolved in very different ways. Although the morphological structures that make up the eye are not homologous, some of the developmental mechanisms involved in patterning eyes certainly are.

Developmental data is currently being rapidly acquired, yet most mechanistic data still do not describe the formation of the complex morphological structures of interest to systematists. Holland (1996) put it nicely:

> Within the genome of every organism is a mine of biological information. They contain the majority of information required for building the body plan from a single cell, the fertilized egg, including instructions for ensuring correct spatial organization and differentiation of specialized cell types. This is manifest through the controlled activation of genes in specific cell types or body regions, at appropriate times of development. Consequently, valuable insight into the mechanisms underlying development can be gained by determining the spatiotemporal patterns of gene transcription in embryos of different development stages. The information carried in the genome goes beyond that required for development, however, it also carries a legacy of evolutionary history. Unfortunately, the field of developmental genetics has not yet reached the stage where complex anatomical or physiological characters can be explained in terms of known genetic cascades. Hence we cannot yet compare the genetic control of homologous characters in most taxa.

Exactly how developmental data can be applied to the interpretation of morphology is unclear. As Doyle (1994) stated, "Although it is perhaps taken for granted that molecular developmental genetics has much to offer systematics, the precise use of developmental information is less clear." He went on to say:

> Molecular systematics as it is currently practiced generally involves sequencing genes and using the resulting nucleotide characters to reconstruct gene trees, from which organismal relationships are then inferred. Future molecular systematic studies may utilize detailed knowledge of the molecular basis of development to hypothesize morphological homologies. This approach would presumably entail comparison of regulation and expression in diverse taxa using batteries of gene and immunological probes to localize the mRNA and protein products of regulatory and target genes in situ. (Doyle 1994)

Along similar lines, DeSalle and Grimaldi (1993) mentioned several morphological characters in *Drosophila* for which the genetic and developmental control might be worked out. They pointed out, "The availability of detailed genetics in *D. melanogaster* indicates the complexity of the genetic control of the morphology." Thus, gathering developmental character data is likely to be a time-consuming and laborious enterprise, relative to the rapidity with which nucleic acid sequences can now be deciphered. New tools and new perspectives have given rise to a plethora of developmental data. And yet the precise connection between phylogenetic systematics and these data remains to be drawn. These data are likely to provide insight into the homology, or the relative level of homology, of different hierarchical levels of the developmental process, from gene expression to morphology. And yet because of dissociation or evolution of subsets of these processes, similarity of development as a homology criterion will not be foolproof.

Ontogeny, Homology, and Terminal Change

As described above, ontogeny is a frequently used criterion in recognizing homology of morphological characters. Its use, however, rests on the assumption that evolutionary change in late ontogeny (terminal change) is more common than change in early ontogeny (nonterminal change). For example, consider a character with observed ontogenetic transformations in two species: $a \rightarrow a$ in one species and $a \rightarrow b$ in the other. Now suppose that the ontogeny is observed in a third species, in which b remains b throughout development $(b \rightarrow b)$ or b transforms into a $(b \rightarrow a)$. To the extent that the development of b from a (i.e., the developmental precedence of a) is used in determining the homology of b, no b in the sequences $b \rightarrow b$ and $b \rightarrow a$ will be recognized as homologous. That is, if *a* is observed to precede *b*, and if the homology of what is observed as *b* among taxa is based on its ontogenetic derivation from *a*, then what one initially thinks of as *b* (which is found to appear in ontogeny without *a*) will be interpreted as nonhomologous with *b*; *b* is actually *c*. Thus, if development is applied as a primary criterion for recognizing homologies among states within developmental sequences (characters), phylogenetic change (both binary and multistate characters) will be interpreted as terminal in ontogeny. Only when other homology criteria are applied, or when characters are interpreted with respect to other characters, can nonterminal change be recognized.

The assumption that evolutionary change most often occurs or is expressed terminally in ontogeny is based on the reasoning that development is hierarchical and that a modification early in ontogeny will result in a large "cascade" of morphological consequences; later modifications will have only minor effects (Horder 1983; Maynard Smith 1983; Arthur 1988). Because the morphological consequences of an early change are presumed to be more detrimental to survival of the organism than changes later in development, it follows that "terminal" change will be the most common form of evolutionary change in ontogeny, although many changes early in ontogeny have been noted (de Beer 1958). This assumption of the conservative nature of early development and the predominance of terminal change is pervasive, appearing in a variety of contexts (Gould 1977; Horder 1983; Nieuwkoop and Sutasurya 1983; Gans 1987; Endress 1990). Maynard Smith (1983), for example, commented, "If development is hierarchical, as most of us would think, then evolutionary changes could occur at any stage. However, in practice it is hard to see how viable alterations could occur except as terminal alterations or additions (if development passes through a free-living larval stage, neotenic changes are also a possibility)." A related reason often given for the presumed commonality of terminal change is that it is "harder" to modify early ontogeny than late ontogeny (Arthur 1982; Freeman 1982; Miyazaki and Mickevich 1982; Cohen and Massey 1983). Gould (1977: 268) wrote, "Since early ontogenetic stages are so refractory to change, evolutionary novelties (if they are to develop gradually) will almost always make their first appearance in late stages." His emphasis on two particular patterns of heterochrony, terminal addition (peramorphosis) and terminal deletion (paedomorphosis), reflects this assumption.

With respect to the ontogeny of particular morphological characters, the expectation that modifications will be expressed later in ontogeny is not incontestable. Consideration of the complexity of developmental processes underlying individual characters, and assuming that such processes might be considered the "character," it is likely that evolutionary change is manifested throughout. In contrast to a simple nested hierarchical set of developmental pathways with a linear correspondence between a gene and its morphological product, nonnested hierarchical pathways such as feedback loops probably underlie most morphological characters (Kauffman 1985; Roth 1988; Nijhout 1990). Although little is known of the developmental mechanisms that underlie most developmental sequences, information indicates that no simple linear relationship, one gene → one character, exists between genes and morphology (Nijhout 1990). It is not at all evident that "terminal" change in the ontogeny of a morphological character should be manifested more often than "nonterminal" change. Data from empirical studies (below), however, suggest that it is.

A total of eight different classes of phylogenetic changes in ontogeny can be distinguished (figure 4.2) through the comparison of ontogenetic change within the context of a phylogeny: terminal addition (peramorphosis), terminal deletion (paedomorphosis), terminal substitution, nonterminal addition, nonterminal deletion, nonterminal substitution, ontogenetic reversal, and substitution (diagrammed in Mabee 1993). In an earlier study (Mabee 1993), I calculated the frequencies of different classes of evolutionary changes in development as interpreted from the outgroup, ontogenetic, and reverse ontogenetic phylogenies for centrarchid fishes (figure 4.2). Basically, these trees represent the best estimate (outgroup tree) and extremes (ontogeny and reverse-ontogeny trees) for terminal addition and terminal deletion. These trees and each character state change and interpretation are discussed in detail in Mabee (1993). On the outgroup tree, 32% of the character state change is interpreted as terminal addition. Terminal addition ranges from 21%, as implied by the reverse-ontogeny tree, to 52%, as implied by the ontogeny tree. About 40% of the character state change is interpreted as terminal deletion on the outgroup hypothesis. Terminal deletion ranges from 24%, as implied by the ontogeny tree, to 53%, as implied by the reverse-ontogeny tree. Frequencies of other classes of character state changes are relatively low.

Terminal modification of character ontogenies is by far the most common type of evolutionary change. Terminal changes (terminal addition, terminal deletion, and terminal substitution) constitute approximately three-fourths of the character state change. Terminal change was found to be the predominant mode of ontogenetic evolution in the other comprehensive empirical study done on the relation between ontogeny and phylogeny (Kraus 1988). Kraus found that all of the changes for the 32 characters could be interpreted as terminal addition and terminal deletion. These studies suggest that early ontogeny is conserved in evolution, with change occurring at the terminal temporal "ends" of ontogenetic sequences. However, as I have pointed out previously (Mabee 1993), interpretation of evolutionary change in ontogeny may be biased by the way that ontogenetic characters are recognized, delimited, and coded.

The effect of character delimitation and coding on interpretation is related to the criteria used to recognize character homology. The situation with binary, presence/absence characters involves homologizing absence, making the assumption that initial absence in ontogeny is homologous with secondary loss.[1] Given that most morphological characters that are present in adults are absent in early ontogeny, relative to an ancestral ontogeny in which absence "transforms" to presence, retained absence in a descendant would be interpreted as terminal deletion. Alternatively, relative to the ancestral ontogeny in which absence is retained, the development of a structure from initial absence in a descendant would be interpreted as terminal addition. Thus, evolution will appear to occur solely through terminal addition and terminal deletion of ontogeny to the extent that characters are conceived with the alternative states of presence and absence, and to the extent that absence precedes presence in ontogeny. Because any systematic character can be restated into the binary terms of presence and absence, evolutionary change might be found "empirically" to consist of only terminal addition and deletion.

PHYLOGENETICS AND FUTURE DIRECTIONS

Ontogeny is not useful in polarizing characters for phylogenetic analysis. The usefulness of ontogeny for ordering characters has not been empirically evaluated and the justification is unclear. Ontogeny, however, can be used as a source of characters for phylogenetic analysis. Because the process of gathering comparative information on developmental mechanisms is time-consuming, however, it is essentially prohibitive if the goal is to expediently gather a sizable number of characters for a phylogenetic analysis. Furthermore, incorporating the transformational component of ontogeny into character hypotheses is not straightforward, and further research into and experimentation with methods of character coding are necessary. The formulation of ontogenetic characters can potentially be tested by means of congruence studies (e.g., Miyamoto and Fitch

[1]The validity of homologizing absence may be questioned. Some would say (e.g., Nelson 1978) that no characters are ever truly lost, they are only modified. However, given that characters can be lost in evolution, it is commonly believed that loss is more easily achieved and occurs more frequently than gain. As such, loss may be more prone to homoplasy and might be weighted less than gain in phylogenetic analyses. Additionally, many of the criteria used to assess homology (similarity in position, structure, etc.) cannot be used on structures that are not present. It is therefore easier for an investigator to make an error in judging homology. One is thus led initially to question any procedure that necessarily involves comparison of absence.

1995), and this is an unexplored area in developmental morphology and systematics.

REFERENCES

Alberch, P. 1985. Problems with the interpretation of developmental sequences. Systematic Zoology 34: 46–58.

Arthur, W. 1982. A developmental approach to the problem of variation in evolutionary rates. Biological Journal of the Linnean Society 18: 243–261.

Arthur, W. 1988. A Theory of the Evolution of Development. John Wiley and Sons, Chichester, U.K.

Brooks, D. R., and E. O. Wiley. 1985. Theories and methods in different approaches to phylogenetic systematics. Cladistics 1: 1–11.

Brummer, G. J. A., C. Hemleben, and M. Spindler. 1987. Ontogeny of extant spinose planktonic foraminifer (Globigerinidae): A concept exemplified by *Globigerinoides sacculifer* (Brady) and *G. ruber* (D'Orbigny). Marine Micropaleontology 12: 357–381.

Bryant, H. N. 1991. The polarization of character transformations in phylogenetic systematics: Role of axiomatic and auxiliary assumptions. Systematic Zoology 40: 433–445.

Burke, A. C., C. E. Nelson, and B. A. Morgan. 1995. *Hox* genes and the evolution of vertebrate axial morphology. Development 121: 333–346.

Cohen, J., and B. D. Massey. 1983. Larvae and the origins of major phyla. Biological Journal of the Linnean Society 19: 321–328.

Cubbage, C. C., and P. M. Mabee. 1996. Development of the cranium and paired fins in the zebrafish *Danio rerio* (Ostariophysi, Cyprinidae). Journal of Morphology 229: 121–160.

de Beer, G. 1958. Embryos and Ancestors. Oxford University Press, London.

De Pinna, M. C. C. 1994. Ontogeny, rooting, and polarity. Pages 157–172 *in* Models in Phylogeny Reconstruction (ed. R. W. Scotland, D. J. Siebert, and D. M. Williams). Clarendon Press, Oxford, U.K.

De Pinna, M. C. C. 1996. A phylogenetic analysis of the Asian catfish families Sisoridae, Akysidae, and Amblycipitidae, with a hypothesis on the relationships of the neotropical Aspredinidae (Teleostei, Ostariophysi). Fieldiana Zoology 84: 1–83.

de Queiroz, K. 1985. The ontogenetic method for determining character polarity and its relevance to phylogenetic systematics. Systematic Zoology 34: 280–299.

DeSalle, R., and D. Grimaldi. 1993. Phylogenetic pattern and developmental process in *Drosophila*. Systematic Biology 42: 458–475.

Doyle, J. J. 1994. Evolution of a plant homeotic multigene family: Toward connecting molecular systematics and molecular developmental genetics. Systematic Biology 43: 307–328.

Endress, P. K. 1990. Patterns of floral construction in ontogeny and phylogeny. Biological Journal of the Linnean Society 39: 153–175.

Fink, W. L., and M. L. Zelditch. 1996. Historical patterns of developmental integration in piranhas. American Zoologist 36: 61–69.

Freeman, G. 1982. What does the comparative study of development tell us about evolution? Pages 155–167 *in* Evolution and Development (ed. J. T. Bonner). Springer-Verlag, New York.

Gans, C. 1987. The neural crest: A spectacular invention. Pages 361–380 *in* Developmental and Evolutionary Aspects of the Neural Crest (ed. P. F. A. Maderson). John Wiley and Sons, New York.

Gould, S. J. 1977. Ontogeny and Phylogeny. Harvard University Press, Cambridge, Mass.

Grande, L. 1985. Recent and fossil clupeomorph fishes, with materials for revision of the subgroups of clupeoids. Bulletin of the American Museum of Natural History 181: 231–372.

Halder, G., P. Callaerts, and W. J. Gehring. 1995. Induction of ectopic eyes by targeted expression of the eyeless gene in *Drosophila*. Science 267: 1788–1792.

Hall, B. K. 1990. Heterochrony in vertebrate development. Seminars in Developmental Biology 1: 237–243.

Hanken, J. 1984. Miniaturization and its effects on cranial morphology in plethodontid salamanders, genus *Thorius* (Amphibia: Plethodontidae). I. Osteological variation. Biological Journal of the Linnean Society 23: 55–75.

Hauser, D., and W. Presch. 1991. The effects of ordered characters on phylogenetic reconstruction. Cladistics 7: 243–265.

Hibbett, D. S., S. Murakami, and A. Tsuneda. 1993. Sporocarp ontogeny in *Panus* (Basidiomycotina): Evolution and classification. American Journal of Botany 80: 1336–1348.

Hillis, D. M. 1995. Approaches for assessing phylogenetic accuracy. Systematic Biology 44: 3–16.

Holland, P. W. H. 1996. Molecular biology of lancelets: Insights into development and evolution. Israel Journal of Zoology 42: S247–S272.

Holland, P. W. H. 1997. Vertebrate evolution: Something fishy about *Hox* genes. Current Biology 7: R570–R572.

Holland, P. W. H., J. Garcia-Fernandez, and N. A. Williams. 1994. Gene duplications and the origins of vertebrate development. Development (supplement): 125–133.

Horder, T. J. 1983. Embryological bases of evolution. Pages 315–352 *in* Development and Evolution (ed. B. C. Goodwin, N. Holder, and C. C. Wylie). Cambridge University Press, Cambridge.

Hull, D. L. 1988. Science as a Process. University of Chicago Press, Chicago.

Judd, D. D. 1998. Exploring component stability using life-stage concordance in sabethine mosquitos (Diptera: Culicidae). Cladistics 14: 63–93.

Kauffman, S. A. 1985. New questions in genetics and evolution. Cladistics 1: 247–265.

Kellogg, E. A., and H. B. Shaffer. 1993. Model organisms in evolutionary studies. Systematic Biology 42: 409–414.

Kitching, I. J. 1992. The determination of character polarity. Pages 22–43 *in* Cladistics: A Practical Course in Systematics (ed. P. L. Forey, C. J. Humphries, I. L. Kitching, R. W. Scotland, D. J. Siebert, and D. M. Williams). Clarendon Press, Oxford, U.K.

Klompen, J. S. H., and B. M. O'Connor. 1989. Ontogenetic patterns and phylogenetic analysis in Acari. Pages 91–103 *in* The Concept of Stase and the Ontogeny of Arthropods (ed. H. M. Andre and J.-C. Lions). AGAR Publishers, Wavre, Belgium.

Kluge, A. G. 1985. Ontogeny and phylogenetic systematics. Cladistics 1: 13–27.

Kluge, A. G. 1988. The characteristics of ontogeny. Pages 57–82 *in* Ontogeny and Systematics (ed. C. J. Humphries). Columbia University Press, New York.

Kluge, A. G., and R. E. Strauss. 1985. Ontogeny and systematics. Annual Review of Ecology and Systematics 16: 247–268.

Kraus, F. 1988. An empirical evaluation of the use of the ontogeny polarization criterion in phylogenetic inference. Systematic Zoology 37: 106–141.

Leis, J. M. 1986. Larval development in four species of Indo-Pacific coral trout *Plectropomus* (Pisces: Serranidae, Epinephelinae), with an analysis of the relationships of the genus. Bulletin of Marine Science 38: 525–552.

Lundberg, J. G. 1973. More on primitiveness, higher level phylogenies, and ontogenetic transformations. Systematic Zoology 22: 327–329.

Mabee, P. M. 1989a. Assumptions underlying the use of ontogenetic sequences for determining character state order. Transactions of the American Fisheries Society 118: 159–166.

Mabee, P. M. 1989b. An empirical rejection of the ontogenetic polarity criterion. Cladistics 5: 409–416.

Mabee, P. M. 1993. Phylogenetic interpretation of ontogenetic change: Sorting out the actual and artefactual in an empirical case study of centrarchid fishes. Zoological Journal of the Linnean Society 107: 175–291.

Mabee, P. M. 1996. Reassessing the ontogenetic criterion: A response to Patterson. Cladistics 12: 169–176.

Mabee, P. M., and J. Humphries. 1993. Coding polymorphic data: Examples from allozymes and ontogeny. Systematic Biology 42: 166–181.

Mabee, P. M., and T. A. Trendler. 1996. Development of the cranium and paired fins in *Betta splendens* (Teleostei: Percomorpha): Intraspecific variation and interspecific comparisons. Journal of Morphology 227: 249–287.

Maddison, W. P., M. J. Donoghue, and D. R. Maddison. 1984. Outgroup analysis and parsimony. Systematic Zoology 33: 83–103.

Maynard Smith, J. 1983. Evolution and development. Pages 33–45 *in* Development and Evolution (ed. B. C. Goodwin, N. Holder, and C. C. Wylie). Cambridge University Press, Cambridge.

Meier, R. 1995. Cladistic analysis of the Sepsidae (Cyclorrhapha: Diptera) based on a comparative scanning electron microscopic study of larvae. Systematic Entomology 20: 99–128.

Meier, R. 1997. A test and review of the empirical performance of the ontogenetic criterion. Systematic Biology 46: 699–721.

Mishler, B. D. 1988. Relationships between ontogeny and phylogeny, with reference to bryophytes. Pages 117–136 *in* Ontogeny and Systematics (ed. C. J. Humphries). Columbia University Press, New York.

Miyamoto, M. M.., and W. M. Fitch. 1995. Testing species phylogenies and phylogenetic methods with congruence. Systematic Biology 44: 64–76.

Miyazaki, J., and M. F. Mickevich. 1982. Evolution of *Chesapecten* and the biogenetic law. Evolutionary Biology 15: 369–409.

Müller, G. B., and G. P. Wagner. 1996. Homology, *Hox* genes, and developmental integration. American Zoologist 36: 4–13.

Nelson, G. J. 1973. The higher-level phylogeny of vertebrates. Systematic Zoology 22: 87–91.

Nelson, G. J. 1978. Ontogeny, phylogeny, paleontology, and the biogenetic law. Systematic Zoology 27: 324–345.

Nelson, G. J. 1985. Outgroups and ontogeny. Cladistics 1: 29–45.

Nelson, G. J., and N. I. Platnick. 1981. Systematics and Biogeography: Cladistics and Vicariance. Columbia University Press, New York.

Nieuwkoop, P. D., and L. A. Sutasurya. 1983. Some problems in the development and evolution of the chordates. Pages 123–135 *in* Development and Evolution (ed. B. C. Goodwin, N. Holder, and C. C. Wylie). Cambridge University Press, Cambridge.

Nijhout, H. F. 1990. Metaphors and the role of genes in development. Bioessays 12: 441–446.

O'Grady, R. T., and G. B. Deets. 1987. Coding multistate characters, with special reference to the use of parasites as characters of their hosts. Systematic Zoology 36: 268–279.

O'Grady, R. T., G. B. Deets, and G. W. Benz. 1989. Additional observations on nonredundant linear coding of multistate characters. Systematic Zoology 38: 54–57.

Patterson, C. 1994. Null or minimal models. Pages 173–192 *in* Models in Phylogeny Reconstruction (ed. R. W. Scotland, D. J. Siebert, and D. M. Williams). Clarendon Press, Oxford, U.K.

Patterson, C. 1996. Comments on Mabee's "Empirical rejection of the ontogenetic polarity criterion." Cladistics 12: 147–167.

Quiring, R., U. Walldorf, U. Kloter, and W. J. Gehring. 1994. Homology of the eyeless gene of *Drosophila* to the small eye gene in mice and in humans. Science 265: 785–789.

Raff, R. A. 1996. The Shape of Life. University of Chicago Press, Chicago.

Raff, R. A., and G. A. Wray. 1989. Heterochrony: Developmental mechanisms and evolutionary results. Journal of Evolutionary Biology 2: 409–434.

Rosen, D. E. 1982. Do current theories of evolution satisfy the basic requirements of explanation? Systematic Zoology 31: 76–85.

Rosen, D. E. 1984. Hierarchies and history. Pages 77–97 *in* Evolutionary Theory: Paths into the Future (ed. J. W. Pollard). John Wiley and Sons, New York.

Roth, V. L. 1988. The biological basis of homology. Pages 1–26 *in* Ontogeny and Systematics (ed. C. J. Humphries). Columbia University Press, New York.

Smith, K. K. 1996. Integration of craniofacial structures during development in mammals. American Zoologist 36: 70–79.

Swofford, D. L., G. J. Olsen, P. J. Waddell, and D. M. Hillis. 1996. Phylogenetic inference. Pages 407–514 *in* Molecular Systematics, 2nd edition (ed. D. M. Hillis, C. Moritz, and B. K. Mable). Sinauer, Sunderland, Mass.

Tyler, J. C., G. D. Johnson, I. Nakamura, and B. B. Collette. 1989. Morphology of *Luvarus imperialis* (Luvaridae), with a phylogenetic analysis of the Acanthuroidei (Pisces). Smithsonian Contributions to Zoology 485: 1–78.

Velhagen, W. A., Jr. 1995. A comparative study of cranial development in the thamnophiine snakes (Serpentes: Colubridae). Doctoral dissertation, Duke University, Durham, N.C.

Velhagen, W. A., Jr. 1997. Analyzing developmental sequences using sequence units. Systematic Biology 46: 204–210.

Wagner, G. P. 1989a. The origin of morphological characters and the biological basis of homology. Evolution 43: 1157–1171.

Wagner, G. P. 1989b. The biological homology concept. Annual Review of Ecology and Systematics 20: 51–69.

Wagner, G. P. 1996. Homologues, natural kinds, and the evolution of modularity. American Zoologist 36: 36–43.

Wake, D. B. 1989. Phylogenetic implications of ontogenetic data. Geobios 12: 369–378.

Wake, D. B., and A. Larson. 1987. Multidimensional analysis of an evolving lineage. Science 238: 42–48.

Weston, P. H. 1988. Indirect and direct methods in systematics. Pages 27–56 *in* Ontogeny and Systematics (ed. C. J. Humphries). Columbia University Press, New York.

Weston, P. H. 1994. Methods for rooting cladistic trees. Pages 125–155 *in* Models in Phylogeny Reconstruction (ed. R. W. Scotland, D. J. Siebert, and D. M. Williams). Clarendon Press, Oxford, U.K.

Wheeler, Q. D. 1990. Ontogeny and character phylogeny. Cladistics 6: 225–268.

Wiens, J. J. 1995. Polymorphic characters in phylogenetic systematics. Systematic Biology 44: 482–500.

Wiley, E. O. 1981. Phylogenetics: The Theory and Practice of Phylogenetic Systematics. John Wiley, New York.

Williams, D. M., R. W. Scotland, and S. Blackmore. 1990. Is there a direct ontogenetic criterion in systematics? Biological Journal of the Linnean Society 39: 99–108.

Wray, G. A., and D. R. McClay. 1989. Molecular heterochronies and heterotopies in early echinoid development. Evolution 43: 803–813.

JOHN J. WIENS

5

CODING MORPHOLOGICAL VARIATION WITHIN SPECIES AND HIGHER TAXA FOR PHYLOGENETIC ANALYSIS

In phylogenetic systematics, the ideal characters are those that vary between the terminal units of an analysis but do not vary within them (Thiele 1993). Of course, in the real world, characters are often less than perfect. Whether the terminal taxa of the analysis are species or some higher taxa that are aggregates of species (e.g., genera, families), it is common to find that many characters vary within these units, and in some cases, these may be the only characters available. For example, among the nine species of the lizard genus *Urosaurus,* 23 of the 24 morphological characters that vary between species vary within one or more species as well (Wiens 1993a). Systematists may use a variety of methods for dealing with this variation, including (1) breaking up the terminal taxon into smaller, invariant units (e.g., breaking up a variable genus into its constituent species), (2) excluding variable characters, and (3) using different methods for inclusion and coding of variable characters. Morphologists only rarely mention how this variation is dealt with in empirical phylogenetic studies (Campbell and Frost 1993; Wiens 1995). Yet the choice among different methods for dealing with variable characters is important, because the application of different methods to the same data can produce very different trees (figure 5.1). The traditional approach for treating within-taxon variation in morphology seems to be the exclusion of these nonfixed characters without mention or justification. However, this topic has recently begun to receive more attention and more explicit treatments (e.g., Campbell and Frost 1993; Wiens 1995; McGuire 1996; Chu 1998; Gutberlet 1998; Poe 1998). In this chapter, I review methods for dealing with characters that vary within terminal taxa, and make recommendations as to what methods may be most effective for morphological data.

Throughout this chapter, I distinguish between variation within species and variation among species within a higher level taxon. I use the term "polymorphism" to refer only to variation within species. This seems to be the most standard usage, although some authors may also refer to interspecific variation as polymorphism (e.g., Nixon and Davis 1991; Donoghue 1994). Polymorphism and interspecific variation in higher taxa are fundamentally different. Polymorphisms evolve via population genetics processes and can be shared between species through common ancestry. In contrast, interspecific variation does not evolve by the same microevolutionary processes (e.g., genetic drift, selection), and is generally due to homoplasy or non-monophyly of the higher taxa (de Queiroz 1987). Although there is some overlap in the phylogenetic methods that are used to code these different types of variation, they are quite different and are treated in separate sections of this chapter.

This chapter is organized as follows. I first discuss the phylogenetic analysis of polymorphism in morphological characters. More specifically, I review methods that have been used to analyze these data, discuss the results of recent studies that compare the relative accuracy of these methods, make recommendations for treating polymorphic characters in empirical phylogenetic studies, and discuss issues in their analysis in need of future research. This same general outline is followed for a discussion of interspecific variation within higher taxa (which derives largely from Wiens 1998b). The main theme of this chapter is that many of the methodologies that are commonly used by morphologists to deal with polymorphism and variation within higher taxa may be relatively inaccurate, and that less-traditional methods, which utilize more information, may give much better results.

Throughout this chapter, I am largely concerned with the question of which methods are most accurate. By accuracy, I mean the similarity between estimated trees and the true phylogeny. In empirical studies the true phylogeny is almost never known, but useful inferences about the accuracy of methods can come from a number of sources, including computer simulations, known

---

→

Figure 5.1. Different methods of coding within-taxon variation give different trees from the same data. Trees are either the strict consensus of multiple equally parsimonious trees or the single shortest tree. **(A)** Trees produced by applying different methods of coding intraspecific variation to the data from Wiens (1993a) for the lizard genus *Urosaurus*. (Figure from Wiens 1995.) **(B)** Trees produced by applying different methods for coding interspecific variation to the data from Estes et al. (1988) for families and other higher taxa of squamate reptiles. (Figure from Wiens 1998a.)

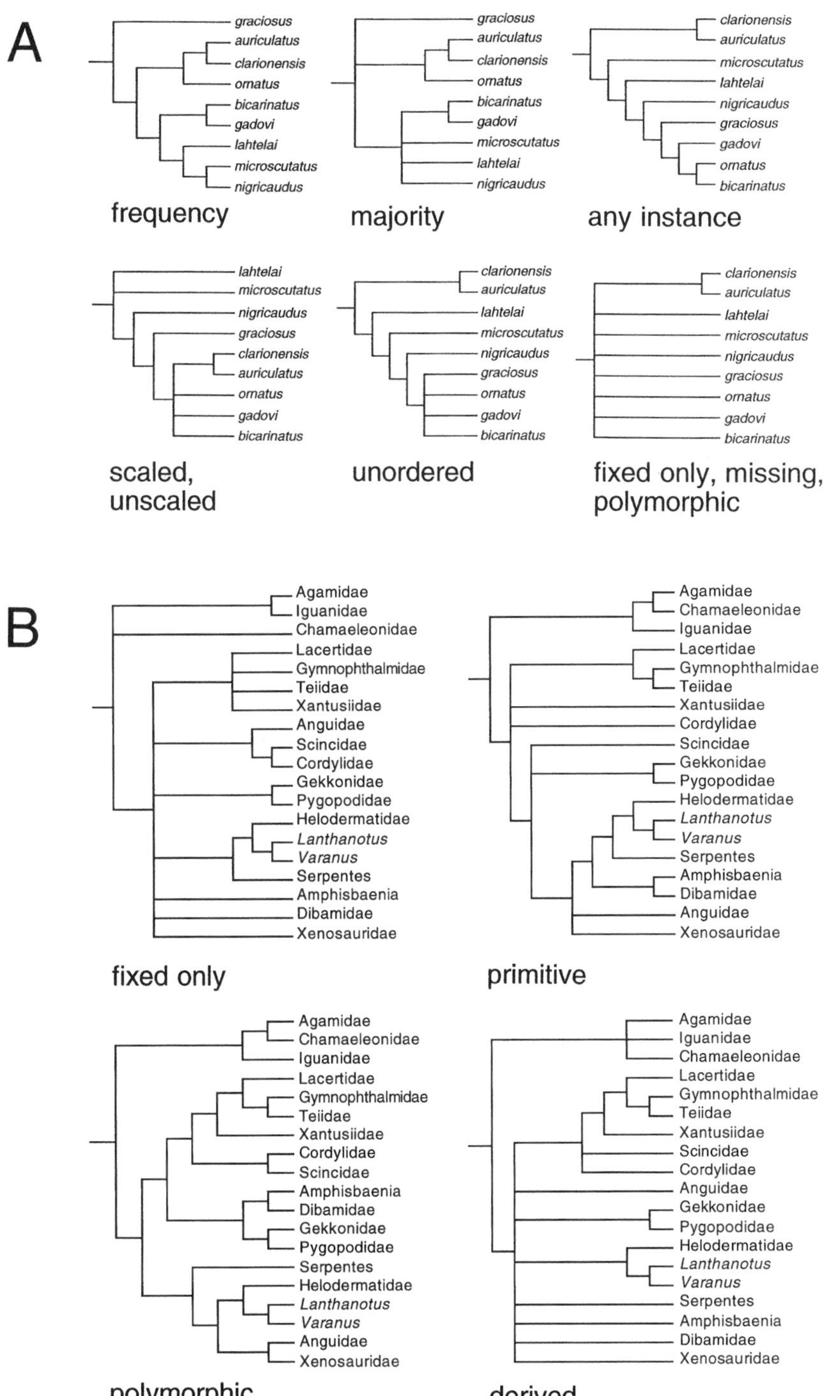
A
graciosus
auriculatus
clarionensis
ornatus
bicarinatus
gadovi
lahtelai
microscutatus
nigricaudus
frequency
graciosus
auriculatus
clarionensis
ornatus
bicarinatus
gadovi
microscutatus
lahtelai
nigricaudus
majority
clarionensis
auriculatus
microscutatus
lahtelai
nigricaudus
graciosus
gadovi
ornatus
bicarinatus
any instance
lahtelai
microscutatus
nigricaudus
graciosus
clarionensis
auriculatus
ornatus
gadovi
bicarinatus
scaled,
unscaled
clarionensis
auriculatus
lahtelai
microscutatus
nigricaudus
graciosus
ornatus
gadovi
bicarinatus
unordered
clarionensis
auriculatus
lahtelai
microscutatus
nigricaudus
graciosus
ornatus
gadovi
bicarinatus
fixed only, missing,
polymorphic
B
Agamidae
Iguanidae
Chamaeleonidae
Lacertidae
Gymnophthalmidae
Teiidae
Xantusiidae
Anguidae
Scincidae
Cordylidae
Gekkonidae
Pygopodidae
Helodermatidae
Lanthanotus
Varanus
Serpentes
Amphisbaenia
Dibamidae
Xenosauridae
fixed only
Agamidae
Chamaeleonidae
Iguanidae
Lacertidae
Gymnophthalmidae
Teiidae
Xantusiidae
Cordylidae
Scincidae
Gekkonidae
Pygopodidae
Helodermatidae
Lanthanotus
Varanus
Serpentes
Amphisbaenia
Dibamidae
Anguidae
Xenosauridae
primitive
Agamidae
Chamaeleonidae
Iguanidae
Lacertidae
Gymnophthalmidae
Teiidae
Xantusiidae
Cordylidae
Scincidae
Amphisbaenia
Dibamidae
Gekkonidae
Pygopodidae
Serpentes
Helodermatidae
Lanthanotus
Varanus
Anguidae
Xenosauridae
polymorphic
Agamidae
Iguanidae
Chamaeleonidae
Lacertidae
Gymnophthalmidae
Teiidae
Xantusiidae
Scincidae
Cordylidae
Anguidae
Gekkonidae
Pygopodidae
Helodermatidae
Lanthanotus
Varanus
Serpentes
Amphisbaenia
Dibamidae
Xenosauridae
derived

(experimental) phylogenies, congruence studies (i.e., well-supported phylogenies), and statistical analyses (see Hillis 1995 for a review). I acknowledge that not all systematists find such studies compelling, and many systematists prefer justifying their methodological choices based on philosophy rather than quantitative analysis (e.g., Brower et al. 1996; Kluge 1997; Siddall and Kluge 1997). Just because these perspectives differ, however, does not necessarily mean that they will disagree in selecting and rejecting phylogenetic methods.

INTRASPECIFIC VARIATION

In this section, I discuss phylogenetic analysis of intraspecifically variable (polymorphic) characters. Before proceeding further, it is useful to distinguish some kinds of intraspecific variation that will not be discussed. Characters that vary only ontogenetically or only between sexes are not considered to be polymorphic; polymorphism is considered to be variation between individuals of comparable age and sex. Furthermore this discussion is limited to discrete or qualitative characters, or characters that are at least described qualitatively (Stevens 1991; Thiele 1993).

The analysis of qualitative, intraspecifically variable morphological characters has generally received very little attention in the systematics literature. Most published morphological studies do not discuss explicitly how polymorphic characters are dealt with, or if they do, it is to say that polymorphic characters were excluded (e.g., see chapter 2 of this book). Furthermore, most conceptual or methodological studies addressing the phylogenetic analysis of intraspecifically variable characters have dealt primarily with analysis of allozyme data (e.g., Mickevich and Mitter 1981, 1983; Swofford and Berlocher 1987; Crother 1990; Swofford and Olsen 1990; Mabee and Humphries 1993; Murphy 1993; Swofford et al. 1996). However, a few recent studies have discussed the phylogenetic analysis of polymorphisms in morphological data (Campbell and Frost 1993; Wiens 1995, 1998b; Wiens and Servedio 1997), and a number of recent morphological studies have begun to use explicit methodologies for dealing with polymorphism (e.g., Campbell and Frost 1993; Wiens 1993a; Domning 1994; McGuire 1996; Reeder and Wiens 1996; Coloma 1997; Chu 1998; Hollingsworth 1998; Poe 1998).

The most practical questions regarding polymorphic characters in morphological phylogenetics are (1) Should these characters be included in phylogenetic analyses or not? and (2) If they are included, what methodologies should be used to analyze them? Different approaches for dealing with polymorphic morphological characters (including their deletion) are reviewed below.

Methods for Treating Intraspecific Variation

In recent years, parsimony-based methods have become the only methods routinely used by morphological systematists. The methods listed below are parsimony methods, but likelihood and distance methods are discussed subsequently. The terminology for most of the parsimony coding methods derives from the reviews of Campbell and Frost (1993) and Wiens (1995), and most of these methods are illustrated in figures 5.2 and 5.3.

Fixed only. With this method, a character is excluded if it exhibits any variation within any species. This method seems to be very widely used in empirical studies (see chapter 2) and has been advocated explicitly by Nixon and Wheeler (1990). The justification for this approach stems from the idea that intraspecifically variable characters are less reliable in inferring phylogeny (e.g., Darwin 1859; Simpson 1961; Farris 1966; Kluge and Farris 1969; Mayr 1969). However, the fixed-only approach is a relatively extreme method. It seems that many morphological systematists include characters that show a limited amount of polymorphism and exclude only the most polymorphic characters (e.g., Campbell and Frost 1993). This seems reasonable, given that polymorphic characters seem to show increasing homoplasy with increasing levels of variability but generally contain useful phylogenetic information (Campbell and Frost 1993; Wiens 1995). Unfortunately, empirical systematists rarely (if ever) state

	Species A	Species B	Species C	Species D
Color of sampled individuals	○○ ○○○	○○ ○●○	●○ ●○●	●● ●●●
Any instance	0	1	1	1
Majority	0	0	1	1
Missing	0	?	?	1
Polymorphic	0	(0,1)	(0,1)	1
Scaled	0	1	1	2
Unordered	0	1	1	2
Unscaled	0	1	1	2

Figure 5.2. Hypothetical example illustrating different methods for coding polymorphic characters. Five individuals of each of four species are sampled; white represents the primitive condition and black is derived.

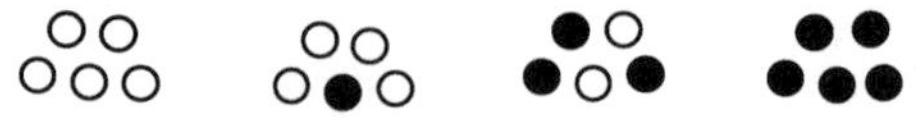

	Species A	Species B	Species C	Species D
Color of sampled individuals	○○ ○○○	○○ ○●○	●○ ●○●	●● ●●●
Proportion dark	0%	20%	60%	100%
Frequency bin	a	f	p	y
Step matrix	A	B	C	D

Frequency bins

Character state	Frequency range (%)
a	0–3
b	4–7
c	8–11
d	12–15
e	16–19
f	20–23
g	24–27
h	28–31
i	32–35
j	36–39
k	40–43
l	44–47
m	48–51
n	52–55
o	56–59
p	60–63
q	64–67
r	68–71
s	72–75
t	76–79
u	80–83
v	84–87
w	88–91
x	92–95
y	96–100

Step matrix

	A	B	C	D
A	0	20	60	100
B	20	0	40	80
C	60	40	0	40
D	100	80	40	0

Figure 5.3. Hypothetical example illustrating two methods for coding polymorphic characters as frequencies. Five individuals of each of four species are sampled; white represents the primitive condition and black is derived. The step matrix represents the different costs (in number of steps) for transitions between each of the character states. The costs are based on the Manhattan distance between the trait frequencies of each species for this character.

a specific criterion for determining exactly how much variation is "too much" for including a polymorphic character (but see Kraus 1988) and do not provide justification for their criteria. Wiens and Servedio (1997) and Wiens (1998b) explored the performance of a number of different exclusion criteria for removing polymorphic characters, based on different levels of intraspecific variability.

Any instance. This parsimony coding method treats a derived trait within a species as present, regardless of whether or not it is fixed or polymorphic, and regardless of the frequency of the trait within a polymorphic species. This method assumes a priori knowledge of the derived state, and it is difficult or impossible to apply to unordered multistate characters. This method was named and proposed by Campbell and Frost (1993) but rejected by them because reversals to the polymorphic state (which might be phylogenetically informative) are hidden.

Majority. Also known as "modal" coding, this method codes a polymorphic species as having the trait that is most common among the sampled individuals. This method, which has the effect of ignoring rare variants within species, may be common in morphological studies, even if it is not stated explicitly.

Missing. With this method, a polymorphic species is coded as unknown ("?") for the variable character. The polymorphic species is treated as if any state could be assigned to the missing cell (regardless of whether or not the state is actually one of those observed in the species), and that cell is uninformative in tree-building (Swofford 1993). A major disadvantage of this method is that the presence of the derived state as a polymorphism cannot be used as evidence to group the polymorphic species with others sharing the derived trait.

Polymorphic. When PAUP or PAUP* (Swofford 1993, 1997) is used, a species coded as polymorphic will be assigned one of the two observed states a posteriori, depending on how the species is placed on the tree by other characters. Thus, the cell coded as polymorphic is largely uninformative in tree-building. This method is equivalent to the missing method when the characters are binary, but with multistate characters coded using the polymorphic method, the variable species will only be assigned one of the states that is actually observed in that species. (This can make certain topological placements of the polymorphic species either more or less parsimonious.) Although this polymorphic coding option is available in PAUP and PAUP*, Swofford (1993) explicitly disavowed its use for coding intraspecific variation.

Scaled. For a binary character, a derived state is coded as being either absent ("0"), polymorphic ("1"), or fixed ("2"). These states are then ordered with the assumption that traits pass through a polymorphic stage between absence and fixed presence. If no polymorphic state is observed for a given character, it is assumed that this state was present and unobserved. This method was named by Campbell and Frost (1993) and is equivalent to the step matrix approach of Mabee and Humphries (1993). The use of a step matrix allows complex ordering of polymorphic multistate characters when there is no clear relationship among the states (as with alleles at an allozyme locus).

Unordered. This method is the same as the scaled method (traits are either absent, polymorphic, or fixed), but the character states are unordered. This method was criticized by Campbell and Frost (1993) and Mabee and Humphries (1993) because no special connection between fixed and polymorphic conditions is implied. Therefore, it is just as easy to evolve a derived trait independently in fixed and polymorphic species as it is to have the trait go from polymorphism to fixation.

Unscaled. This method is the same as the scaled method, except that characters in which no polymorphism is observed are not assumed to have passed through a polymorphic state. Thus, fixed characters effectively receive a weight of 1 with the unscaled method and a weight of 2 with the scaled method (relative to polymorphic characters). This was the method preferred by Campbell and Frost (1993).

Frequency. Frequency parsimony methods make explicit use of the frequency of traits within polymorphic species, giving little weight to small differences in frequencies between species and greater weight to larger differences. This can be implemented in a number of ways, each with different advantages and disadvantages. The most precise parsimony method is implemented in the FREQPARS program (Swofford and Berlocher 1987), which uses trait frequencies directly. However, FREQPARS has a very weak tree-searching algorithm and is therefore impractical for all but the smallest of data sets. I have used a method (suggested by D. Hillis) that approximates the use of FREQPARS but still allows for powerful tree searches (Wiens 1995; see figure 5.3). This method involves giving each taxon a unique character state and then weighting the changes between these states based on the Manhattan distance (Swofford and Berlocher 1987) between the trait frequencies of each pair of species for that character. (See Berlocher and Swofford 1997 for a more detailed discussion.) The main disadvantage of this approach (beside the fact that ancestral states can only be sets of frequencies that are observed in extant species) is that it uses step matrixes, which can slow down tree searches prohibitively if large numbers of taxa are being analyzed. A third method uses ordered frequency bins (defined as ranges of frequencies of a given trait) as character states (e.g., state $a = 0–3\%$, $b = 4–7\%$, etc.; figure 5.3). This frequency-bins method (Wiens 1993a, 1995) is the least precise of the three frequency methods and is designed for binary characters only. However, it is practical for large numbers of taxa (>100; Wiens and Reeder 1997), allows calculations of the fit of characters to the tree, and does allow ancestral states to be reconstructed that are not present in the extant taxa.

The most contentious issue in the use of frequency methods is not what method for using frequencies is best, but rather whether or not frequency information should be used at all. Many authors (e.g., Mickevich and Johnson 1976; Farris 1981; Mickevich and Mitter 1981, 1983; Crother 1990; Campbell and Frost 1993; Mabee and Humphries 1993) have argued that frequencies may not be phylogenetically informative because of their variability within species over space and time. In contrast, Swofford and Berlocher (1987) argued that qualitative methods that ignore frequencies are too subject to sampling error. They argued that trait absence and trait presence can be distinguished reliably only with extremely large sample sizes and that the sampling and detection of rare variants among species within a clade may not reflect the actual distribution of the traits or the phylogeny of the group. Qualitative methods give equal weight to rare and common traits within a species, and thus may be misled, whereas frequency methods give little weight to rare traits and are therefore unlikely to be problematic in this way. I found (Wiens 1995) that frequency-coded polymorphic characters (both allozyme and morphological) do contain significant phylogenetic information and are relatively insensitive to variation in sample size, and that qualitative coding methods (any instance, scaled, unordered, unscaled) seem to show increasing levels of noise with increasing sample size (as predicted by Swofford and Berlocher 1987) and may completely randomize the phylogenetic information present in a data set.

Confidence coding. Domning (1994) has developed a unique method which statistically incorporates sample size when coding polymorphic characters. I refer to this approach as confidence coding. It involves finding the frequency of traits within each species and the 95% confidence interval for each of these frequencies, based on the sample size. For a given species, if the lower confidence limit of the frequency of the commonest trait is greater than 0.5, the species is coded as having the majority condition. If not (or if two traits are present at equal frequencies), the taxon is coded "whichever way was more congruent with other characters (i.e., whichever way did not imply a reversal)"(Domning 1994). The congruence with other characters is determined on the basis of a preliminary tree. This method was not included in studies comparing method performance (e.g., Wiens 1995, 1998b; Wiens and Servedio 1997, 1998) but would likely perform similarly to the majority method.

Individuals as terminals. Some authors have advocated coding individual organisms as terminal units in phylogenetic analyses (e.g., Vrana and Wheeler 1992) as a general solution for dealing with intraspecific variation. Although this approach is widely used in molecular studies, it has never been applied (to my knowledge) in morphological phylogenetics. A serious disadvantage of this approach is that it does not allow polymorphisms shared between species to act as synapomorphies (Wiens 1998a). Instead all polymorphisms will either (1) cause some of the individuals of different species to cluster together on the basis of their shared polymorphisms, thus making the species appear as non-monophyletic (regardless of whether or not they are actually distinct species), or (2) be treated as homoplasies. It is true that polymorphism in a given species may actually indicate the presence of more than one species (or it may indicate that two species are really only one), and it is true that polymorphisms may be present in two species because of homoplasy rather than shared ancestry. However, an approach that tends to maximize homoplasy a priori is obviously problematic. Nevertheless, the accuracy of the individuals-as-terminals method relative to other coding methods has not yet been tested and should be addressed.

Character weighting. Campbell and Frost (1993) and I (Wiens 1995) have advocated character weighting as a means to include polymorphic characters while accommodating their higher levels of homoplasy. Campbell and Frost (1993) recommended successive weighting (Farris 1969) coupled with the unscaled coding method. I (Wiens 1995) advocated a priori down-weighting based on levels of intraspecific variability (Farris 1966).

Distance and likelihood methods. Distance methods (e.g., UPGMA, neighbor-joining, Fitch-Margoliash) are often applied to polymorphism in allozyme data, and Felsenstein's (1981) continuous maximum likelihood method, designed explicitly for phylogenetic analysis of allele frequency data, is sometimes used as well. Distance and likelihood methods are rarely (if ever) applied to polymorphic morphological data (Wiens 1998b). These methods are intended for allele frequency data (i.e., allozymes), but they can be applied to morphological characters by treating the frequencies of morphological traits within a species as if they were alleles at an allozyme locus (Wiens 1998b). The potential advantage of likelihood and (at least some) distance methods is that they may gave accurate estimates of phylogeny under conditions where parsimony is thought to fail (e.g., Felsenstein 1978; Huelsenbeck 1995).

Testing the Accuracy of Phylogenetic Methods

The accuracy of most of the methods listed above has recently been tested using computer simulations (Wiens and Servedio 1997, 1998) and congruence analysis of morphological data (Wiens 1998b). Simulations have two main benefits: (1) the true phylogeny is known unequivocally, and (2) the parameters that may affect the performance of methods can be varied systematically and thereby understood (Hillis 1995; Huelsenbeck 1995). The disadvantage of simulations is that they always invoke a number of simplifying assumptions. These assumptions may bias the results, and the simplicity of the simulated conditions may make them inapplicable to phylogenetic problems in the real world (Hillis 1995; Huelsenbeck 1995).

Congruence studies offer a more realistic alternative to simulations. With congruence analysis, the ability of methods to recover clades that are well supported by multiple data sets with a limited sample of characters is tested (Miyamoto and Fitch 1995). Although the true phylogeny is not known in such cases (as it is in simulations), the best explanation for clades that are well supported by multiple data sets is that they represent real evolutionary groups (Miyamoto and Fitch 1995). The congruence approach assumes that some clades are effectively known but uses data that are generated by real evolutionary processes (Miyamoto and Fitch 1995). However, these processes are largely unknown, and parameters generally cannot be manipulated as they can in simulations (for example, tree shape, branch lengths, and number of states per character). In summary, both simulations and congruence analyses make strong but non-overlapping assumptions, and they are complementary approaches that can be used together to test the accuracy of methods (Allard and Miyamoto 1992).

Results of Simulation and Congruence Analyses

In general, the results of the simulation and congruence analyses of polymorphic mophological characters are remarkably similar, despite the many simplifying assumptions associated with the simulations. Figure 5.4 shows a sample of the results from the simulation and congruence analyses, with parameters matched as closely as possible. The major conclusions of the simulations and congruence analyses are the following.

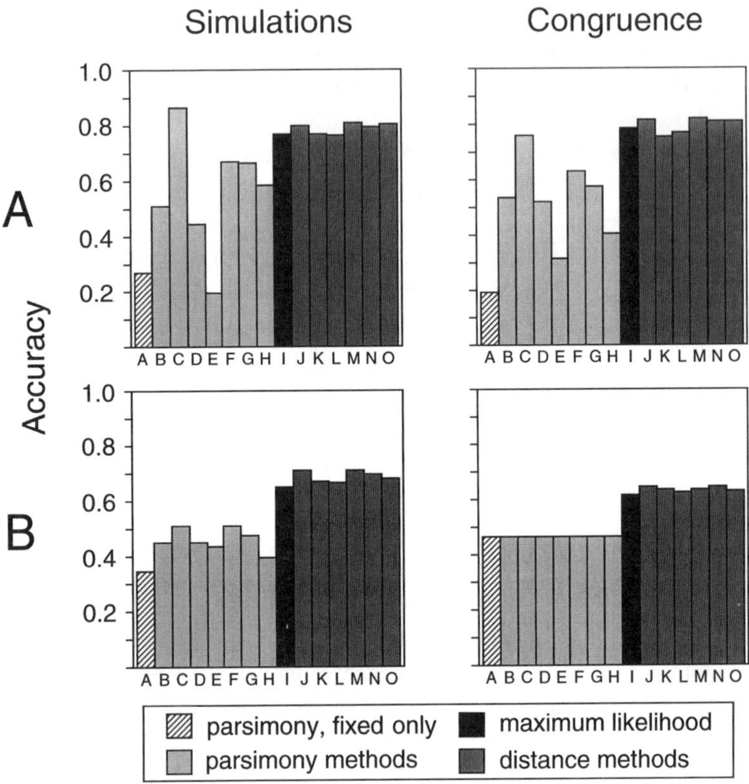

Figure 5.4. Accuracy of methods for analyzing polymorphic data. Sample of results from simulations (Wiens and Servedio 1997, 1998) and congruence analyses (Wiens 1998b), for eight taxa and 25 characters per data set. Results from congruence analyses are for *Sceloporus*. The simulations results are from data sets in which branch lengths were varied randomly among lineages (from 0.2 to 2.0). Each bar represents the average accuracy from 100 replicated matrixes, and accuracy is the number of nodes in common between the true and estimated phylogenies. Parsimony methods: A, fixed only; B, any instance; C, frequency (step matrix); D, majority; E, missing and polymorphic; F, scaled; G, unordered; H, unscaled. Maximum likelihood and distance methods: I, continuous maximum likelihood; J, UPGMA with Nei's (1972) distance; K, neighbor-joining with Nei's distance; L, Fitch-Margoliash with Nei's distance; M, UPGMA with Cavalli-Sforza and Edwards's (CSE; 1967) modified chord distance; N, neighbor- joining with CSE chord distance; O, Fitch-Margoliash with CSE chord distance. **(A)** Sample size of 10 individuals per species for simulations and close to 10 for congruence analyses (average of 11 individuals per species for external characters and 3 individuals per species for osteological characters). **(B)** One individual sampled per species. When $n = 1$, the parsimony methods give identical results in the congruence analyses because heterozygotes cannot be detected (so no polymorphism is observed), whereas heterozygotes can be detected in the simulations.

(1) The fixed-only approach (i.e., excluding all polymorphic characters) consistently decreases accuracy (see figure 5.4).

(2) Excluding some polymorphic characters (using the various exclusion criteria) and downweighting polymorphic characters (with a priori or successive-weighting) may improve accuracy under some conditions, but they do not improve accuracy consistently relative to the unweighted-frequency parsimony method including all polymorphic characters.

(3) The frequency coding method appears to be the most generally accurate parsimony method (e.g., figure 5.4), followed by the scaled and unscaled methods (recommended by Mabee and Humphries [1993] and Campbell and Frost [1993], respectively). The relative performance of the various parsimony coding methods seems to reflect the amount of information from polymorphic characters that each method utilizes. The frequency method uses the most fine-grained information and generally performs best. The scaled, unscaled, and unordered methods recognize at least three states for polymorphic characters (absent, polymorphic [frequencies from 1 to 99%], and fixed), and are exceeded in accuracy only by the frequency method. The any-instance and majority approaches recognize only two possible character states and generally fare worse than the preceding methods. The missing and polymorphic methods effectively treat polymorphisms as empty data cells (unknowns) and generally perform the worst among the coding methods. Superficially, the frequency method may appear superior only because it provides greater resolution. Although the frequency method does tend to provide more well-resolved trees (Wiens 1995), it is clear that this resolution tends to be accurate. Furthermore, the scaled and frequency methods consistently pick different trees in real data sets (Wiens 1995), which suggests that differences in accuracy are not merely due to high versus low resolution of trees. Simulations indicate that some of the differences in accuracy between frequency and non-frequency methods may be related to the greater tendency of non-frequency methods to be misled by unequal branch lengths (Wiens and Servedio 1997, 1998).

The simulation and congruence results (and the randomization tests of Wiens 1995) suggest that frequencies do indeed contain useful phylogenetic information, despite the fact that they can vary greatly within species over space and time. Population genetics theory (e.g., Kimura 1955) shows that when the frequency of a neutral polymorphic trait is modified extensively within a species, the trait is likely to become fixed or lost. As long as a trait with a high frequency within a species tends to go to trait fixation and a trait with low frequency tends to be lost (Kimura 1955), then frequencies should be informative, even when they change rapidly within species. Although there are factors that might cause trait frequencies to fluctuate dramatically within species without going to loss or fixation (i.e., frequency-dependent selection, temporally varying selection), these forces seem very unlikely to be operating in a large proportion of systematic characters within many species.

(4) Small sample sizes (e.g., $n = 1$ individual per species) may decrease accuracy greatly, particularly when levels of polymorphism are high (see figure 5.4). For example, simulations with eight species, two alleles per locus, 25 characters, and a length of 0.2 (95% characters known to be polymorphic in one or more species) show that the frequency method has an accuracy of 74% with 10 individuals per species and an accuracy of 18% with 1 individual per species (Wiens and Servedio 1997). For comparable conditions, congruence analyses with data from *Sceloporus* (92% of the characters observed as polymorphic in one or more species) found that the frequency parsimony method had an accuracy of nearly 80% with full sample sizes (about 11 individuals per species for external characters and 3 per species for osteological characters) and an accuracy of 46% with a sample size of 1 (figure 5.4). Simulations (Wiens and Servedio 1997) suggest that when levels of polymorphism are high, accuracy may be increased more efficiently by increasing the number of individuals sampled per species than by increasing the number of characters.

(5) Distance and likelihood methods (including UPGMA) often outperform parsimony methods on real and simulated morphological data sets (see figure 5.4). The distance and likelihood methods use fine-grained information on trait frequencies, and they perform similarly to the frequency parsimony method under many conditions. However, the distance and likelihood methods consistently outperform parsimony when sample sizes are small (e.g., $n = 1$ individual per species). This seems to be related to the tendency of distance and likelihood methods to give fully resolved estimates and to the superior ability of these methods to cope with random noise (Wiens 1998b).

In addition to having apparent superiority at small sample sizes, likelihood and additive distance methods (neighbor-joining and Fitch-Margoliash, but not UPGMA) consistently outperform parsimony in simulations under certain combinations of branch lengths among taxa (Wiens and Servedio 1998). Specifically, when there are unrelated terminal lineages with long branches (i.e., small population sizes under the genetic drift model simulated) separated by a short internal branch (e.g., see figure 5.5), the additive distance methods and likelihood will tend to be accurate, whereas parsimony and UPGMA will tend to incorrectly place the taxa with long branches together as sister taxa, even when a huge number of characters has been sampled. These results suggest that the Felsenstein Zone effect, which has been studied in detail for fixed characters (e.g., Felsenstein 1978; Huelsenbeck 1995), may also apply to polymorphic data (see also Kim and Burgman 1988). These results are particularly interesting because likelihood and additive distance methods currently are almost never applied to morphological data.

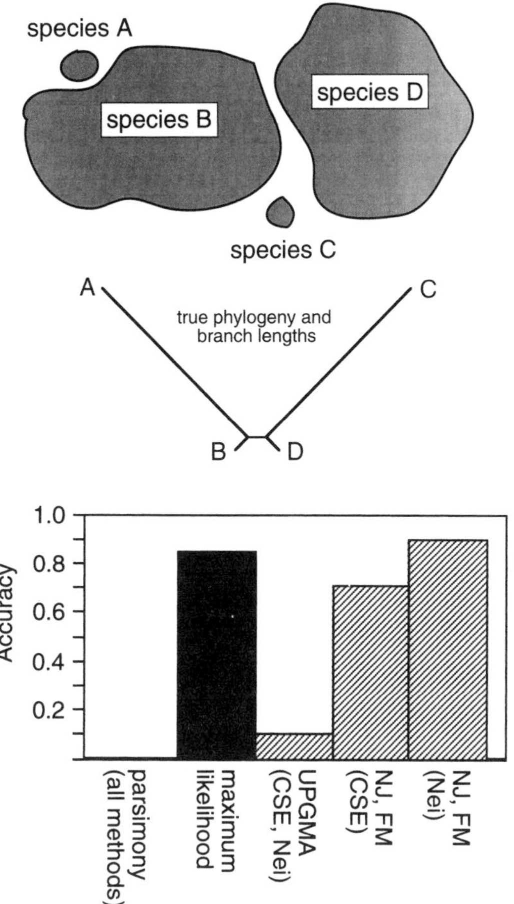

Figure 5.5. Hypothetical example showing the Felsenstein Zone effect for polymorphic data. Species A and C have small geographic distributions and small population sizes, and thus have long branches under a genetic drift model of evolutionary change. Species B and D have have large distributions, large population sizes, and short branch lengths. Simulations (Wiens and Servedio 1998) suggest that, under these conditions, all parsimony methods and UPGMA will tend to incorrectly put A and C together on the basis of shared fixations, losses, and changes in trait frequency. In contrast, continuous maximum likelihood and the additive distance methods—neighbor-joining (NJ) and Fitch-Margoliash (FM) methods with either Nei's (1972) distance or Cavalli-Sforza and Edwards's (CSE; 1967) modified chord distance—can give accurate results under these conditions, given enough characters. Each bar represents the proportion of 100 replicated matrixes in which the true phylogeny is correctly resolved. Simulation results are for long branch lengths of 1.9, short branch lengths of 0.1, 500 characters and four species per data set, and trait frequencies within species sampled without error. See Wiens and Servedio (1998) for further details.

Objections to Using Frequency Data in Morphological Phylogenetics

Recent statistical analyses (Wiens 1995), simulation studies (Wiens and Servedio 1997, 1998), and congruence analyses (Wiens 1998a) suggest that methods that incorporate polymorphic characters and frequency information will tend to give the most accurate estimates of phylogeny. A number of recent morphological studies have used the frequency-bins method to code polymorphic characters (e.g., Wiens 1993a; McGuire 1996; Reeder and Wiens 1996; Coloma 1997; Wiens and Reeder 1997; Chu 1998; Gutberlet 1998; Hollingsworth 1998; Poe 1998; Grismer 1999; Harvey and Gutberlet 1999). Nevertheless, the use of frequency information in phylogenetic analysis remains controversial (e.g., Siddall and Kluge 1997). Objections to the use of frequency information include claims that (1) frequencies are not heritable, organismal traits, (2) large sample sizes are required for frequency methods to be useable, and (3) frequencies vary too much within species to be phylogenetically informative. I address these below.

It is true that frequencies are neither heritable nor organismal traits. By nonheritable, I mean that nonfixed frequencies are probably not passed from ancestor species A to descendant species B without at least some change (but this applies to quantitative characters as well, such as the length of a bone). From a practical perspective, it is also clear that lack of heritability does not seem to impede the ability of frequency methods to accurately estimate phylogeny. As long as traits that are present at high frequencies tend to go to fixation rather than loss (and vice versa), frequency methods perform well and may be more accurate than non-frequency methods, even when frequencies at nonfixed loci are effectively randomized by extensive evolutionary change (Wiens and Servedio 1997). The philosophical objection appears to be hollow as well. It is true that frequencies are traits of taxa, and not of individual organisms. However, this is true for polymorphisms in general. Furthermore, many morphological characters have states that implicitly or explicitly describe a range of values (e.g., coding a structure as "long" versus "short"), and the states therefore pertain to species rather than individuals, just like frequencies (Thiele 1993). Surprisingly, although Siddall and Kluge (1997) objected to frequencies because they are not organismal traits, Kluge has used non-organismal traits in his empirical studies, such as coding of the modal condition within a species (e.g., Kluge 1993). (Modes are neither organismal nor heritable.) In fact, exclusion of polymorphic characters and frequency information is contrary to the maxim of total evidence (Kluge 1989), which requires that all the relevant data be included and explained.

An objection to frequency methods that is sometimes raised in specific cases is that sample sizes (individuals per species) may be insufficient for this type of coding method (e.g., Campbell and Frost 1993). The implicit assumption in this case seems to be that, at small sample sizes, frequency-based methods will be less accurate than qualitative coding methods. This appears to be untrue. Simulations suggest that as sample sizes decrease, the performance of methods becomes increasingly similar. This result makes intuitive sense, because if there were no polymorphism, all the coding methods would be identical. However, it should be noted that even with small sample sizes (e.g., $n = 1$ or 2 individuals per species), there are still differences in accuracy among methods, and frequency methods may still outperform other coding methods (see figure 5.4).

The most common objection to frequency data appears to be that frequencies are too variable within species over space and time to be used in phylogenetic analyses. This is an empirical prediction that has little or no empirical support. Many authors have cited the study by Crother (1990) as evidence for the inapplicability of frequency information in phylogenetic analysis (e.g., Campbell and Frost 1993; Mabee and Humphries 1993; Mink and Sites 1996). However, Crother's study had many problems that were not mentioned by these authors. Crother analyzed allele frequency data from four populations of *Microtus ochrogaster* and found that phylogeny estimates based on the same locus differed from year to year. Crother (1990) then concluded that frequencies vary too much over time and space to be phylogenetically informative for reconstructing relationships among species. However, the "populations" were not natural populations from different localities, but rather individuals drawn from the same locality confined in four enclosures (Gaines et al. 1978). Thus, there was no true phylogeny to be estimated for these populations. Furthermore, the frequency method used was that of Swofford and Berlocher (1987), who stated explicitly that their method should not be used to reconstruct phylogenies within species. None of these points were acknowledged by Crother (1990) or subsequent authors, yet Crother's results have been uncritically extrapolated to the interspecific case and generalized to include all applications of frequency data. In contrast, I found (using the g_1 index of Hillis 1991) significant phylogenetic signal in several frequency-coded interspecific data sets (Wiens 1995), and subsequent authors have also found significant phylogenetic signal in their frequency-coded morphological data sets (e.g., McGuire 1996; Wiens and Reeder 1997; Poe 1998; plus my analyses of the data of Chu [1998] and Hollingsworth [1998]). The results of Crother (1990) do show that frequency methods can yield a fully resolved tree despite the absence of an underlying phylogeny (Wiens 1995), but bootstrapping indicates that these resolutions are weakly supported (bootstrap values < 50%). Furthermore, analysis of tree-length distribution skewness (g_1 index) for these data confirms their lack of phylogenetic structure (pers. obs.). The latter result also shows that g_1 analysis can detect data sets that lack underlying phylogenetic structure and that g_1 analysis is not invalidated by frequency coding. It should be understood that I am not claiming that frequencies of traits are not variable within species over space and time. Clearly they can be, but the stability of trait frequencies within species is not necessary for frequency methods to be accurate. One reason for this is that when the frequency of a trait is modified extensively, there is a high probability the trait will become lost or fixed (Kimura 1955). Losses and fixations are phylogenetically informative for frequency methods and may prevent further oscillation of trait frequencies (Wiens and Servedio 1997). As long as traits at high frequencies tend to be fixed rather than lost (and traits at low frequencies tend to be lost), then frequency methods are likely to be advantageous relative to qualitative coding methods.

Areas for Future Research

Although there appears to be considerable justification for applying frequency-based methods to polymorphic data, there is also need for further study. Simulations of polymorphic data have so far been based on relatively simple models, and many realistic complexities have not been included (e.g., high mutation rates, selection, geographic variation, differences between genotype and phenotype, multiple loci affecting traits, larger numbers of taxa). The congruence analyses suggest that the simulation results are similar to those obtained for real data. However, the congruence analyses were based on only one group of lizards and should be tested more widely. In fact, most of the applications of frequency coding to morphology have so far come from vertebrates (especially lizards), and statistical and congruence analyses should be performed on other groups as well, particularly plants and invertebrates. Additional testing of maximum likelihood and the additive-distance methods is of particular concern. The simulation and congruence analyses suggest that these methods often outperform parsimony (especially when there is long-branch attraction), and yet these methods are almost never applied to morphological data. There is also the need to make these distance and likelihood methods more easily applicable to morphological data. As implemented in PHYLIP (Felsenstein 1995), these methods have a number of practical limitations that make them difficult to use on many data sets (e.g., no missing data allowed, no combination of different data types possible). Furthermore, the distance and likelihood methods discussed here were designed with allozyme data in mind (and are based on very simple models), yet they performed well, relative to parsimony, on morphological data (Wiens 1998b). Methods that are specifically tailored for use with polymorphic morphological data should perform even better.

Conclusions and Recommendations

The results from simulations and congruence analyses strongly suggest that excluding polymorphic characters decreases phylogenetic accuracy and that the best methods for analyzing polymorphic data are those that make direct use of frequency information. However, very few published empirical morphological studies utilize both polymorphic characters and information on trait frequencies. Thus, these results would seem to support major changes in the practices morphological systematists use to deal with intraspecific variation. Because the empirical data supporting these conclusions is limited, further quantitative testing of polymorphic characters and of methods for analyzing them is clearly in order. However, further analyses will be difficult until more morphological systematists report data on polymorphic characters and the frequencies of traits within species.

INTERSPECIFIC VARIATION

Many phylogenetic analyses of morphological data address the relationships of higher taxa and have supraspecific taxa as their terminal units (e.g., Gauthier et al. 1988; Rowe 1988; Frost and Etheridge 1989; Trueb and Cloutier 1991; Eernisse et al. 1992; Novacek 1992; Wiens 1993b; Schultze 1994; Carlson 1995; Livezey 1996, 1997; Smith 1996; Rodriguez 1997; Shaffer et al. 1997; Whiting et al. 1997). Using genera, families, or higher ranking groups as terminals requires coding variation among the species that make up these taxa. A number of methods have been used to deal with this variation, such as coding inferred ancestral states, excluding the variable characters, coding variable taxa as missing, and dividing the variable taxa into smaller, monomorphic, taxonomic units. Several authors have provided useful reviews of this problem (e.g., de Queiroz 1987; Estes et al. 1988; Nixon and Davis 1991; Donoghue 1994; Mishler 1994; Yeates 1995; Rice et al. 1997; Bininda-Emonds et al. 1998). In this section, I briefly review the different methods for treating interspecific variation in higher taxa, discuss the results of a recent simulation study in which the accuracies of these methods were compared (Wiens 1998a), make recommendations for empirical studies, and discuss problems in need of further research.

Methods for Treating Interspecific Variation

Fixed only. By this approach, characters are excluded if there is any or too much variation among the species within the higher-level terminal taxa. This

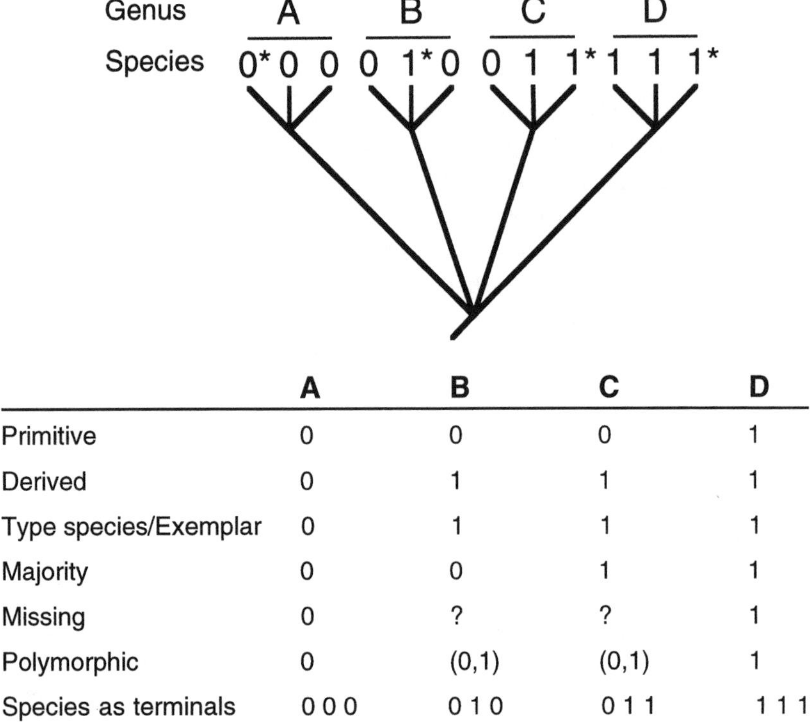

	A	B	C	D
Primitive	0	0	0	1
Derived	0	1	1	1
Type species/Exemplar	0	1	1	1
Majority	0	0	1	1
Missing	0	?	?	1
Polymorphic	0	(0,1)	(0,1)	1
Species as terminals	0 0 0	0 1 0	0 1 1	1 1 1

Figure 5.6. Hypothetical example illustrating different methods of coding interspecific variation in higher taxa. State 0 is primitive and state 1 is derived. There are four genera with three species each, and the asterisks indicate the type species (or exemplar) of each genus.

approach appears to be commonly used, because interspecific variation within terminal taxa is a frequent criterion for excluding characters (see chapter 2).

The following methods involve coding variable characters and are illustrated in figure 5.6.

Splitting up terminal taxa (species as terminals). The practice of splitting up a variable terminal taxon into its constituent units (e.g., species) has several advantages, including (1) providing a potential test of the monophyly of the variable higher taxon (Nixon and Davis 1991), (2) avoiding arbitrary coding

(Nixon and Davis 1991; Yeates 1995), and (3) (implicitly) inferring the ancestral state of the higher taxon in the course of a global parsimony analysis. The potential disadvantage of this approach is that it could lead to analyzing hundreds of taxa simultaneously (which may be computationally difficult), although this depends on how thoroughly the higher taxon is sampled.

The next six methods involve assigning a single character state to the polymorphic terminal taxon. All of these methods implicitly assume that the terminal taxon is known to be monophyletic.

Inferring the ancestral state from the phylogeny within the terminal taxon (IAS). If higher taxa are used as terminals, then the goal of coding these taxa is presumably to represent the character states present in the ancestral species of each group, because these terminal taxa represent the common ancestor of each group. Given this, a number of authors have recommended using information on the phylogeny within the terminal taxon (if available from previous or independent studies) to estimate the ancestral state within each terminal (e.g., Estes et al. 1988; Gauthier et al. 1988; Rowe 1988; Frost and Etheridge 1989; Trueb and Cloutier 1991; Wiens 1993b; Schultze 1994; Livezey 1996). This widespread methodology has been called the placeholder approach (Donoghue 1994), compartmentalization (Mishler 1994), intuitive groundplan method (Yeates 1995), inferred ancestral states (IAS; Rice et al. 1997), and the ancestral method (Bininda-Emonds et al. 1998). I prefer the shortest and most descriptive of these names, IAS.

In many cases, information on the phylogeny within the polymorphic terminal taxon is questionable or unavailable. At least five methods have been proposed for these cases, which are described below.

Primitive state. This method involves coding the polymorphic taxon as having the plesiomorphic state determined in the higher level analysis (i.e., given that higher level taxa make up the ingroup, then the state present in the outgroups of these higher taxa).

Derived state. This method is merely the opposite of the preceding method; rather than coding the polymorphic taxon with the primitive state determined from the higher level outgroup analysis, one codes it with the derived state.

Type species (or single exemplar). Walker et al. (1990) advocated coding a higher taxon based on the taxon's name-bearing type, and many authors code higher taxa based on data from a single representative species.

Majority. This method involves coding higher taxa based on the modal or most common condition among the species (e.g., Livezey 1986; Trueb and Cloutier 1991), and it rests implicitly on the assumption that "common equals primitive." This was referred to as the "democratic" method by Bininda-Emonds et al. (1998).

Missing and polymorphic. Many systematists code variable higher taxa as being polymorphic (having both states) or unknown (missing data), particularly when information on the phylogeny within the variable terminal taxon is unavailable or gives ambiguous reconstructions (e.g., Frost and Etheridge 1989; Livezey 1996). As in the intraspecific case, the main disadvantage of these methods is that cells coded as missing or polymorphic in the data matrix are largely uninformative in reconstructing the tree. Variable taxa are treated as having the state that is most parsimonious given the position of the taxon on the tree, but that position is determined largely by other characters. However, for multistate characters coded by the polymorphic method, the observed states can constrain the placement of the variable taxon somewhat (e.g., if the observed states in a variable higher taxon are 1 and 2, placing that taxon in a clade having state 0 will be more costly than placing it in a clade with state 1 or state 2). In contrast, data cells coded as missing are treated as if any state is possible (so the placement of taxa is completely unconstrained). The missing and polymorphic coding methods give identical results for binary characters.

Testing the Accuracy of Methods with Simulations

I used simulations to compare the accuracy of most of these methods (Wiens 1998a). The simulated data sets included six higher taxa with 3–21 species each (42–66 species total), and 100–400 binary characters evolving at various rates (branch lengths). Readers are referred to Wiens (1998a) for details of the simulations.

The general conclusions from these simulations (figure 5.7) are (1) the commonly used fixed-only method (excluding characters that vary within higher taxa) performs very poorly; (2) the species-as-terminals method is the most accurate under almost all conditions and is often superior to the other methods by a large margin; (3) coding higher taxa based on a single species usually gives relatively poor results, even when many characters are sampled; and (4) among the different methods for coding variable characters, the majority method generally performs best.

The simulation results strongly suggest that higher taxa should be coded by using their constituent species as terminal taxa, rather than the widespread prac-

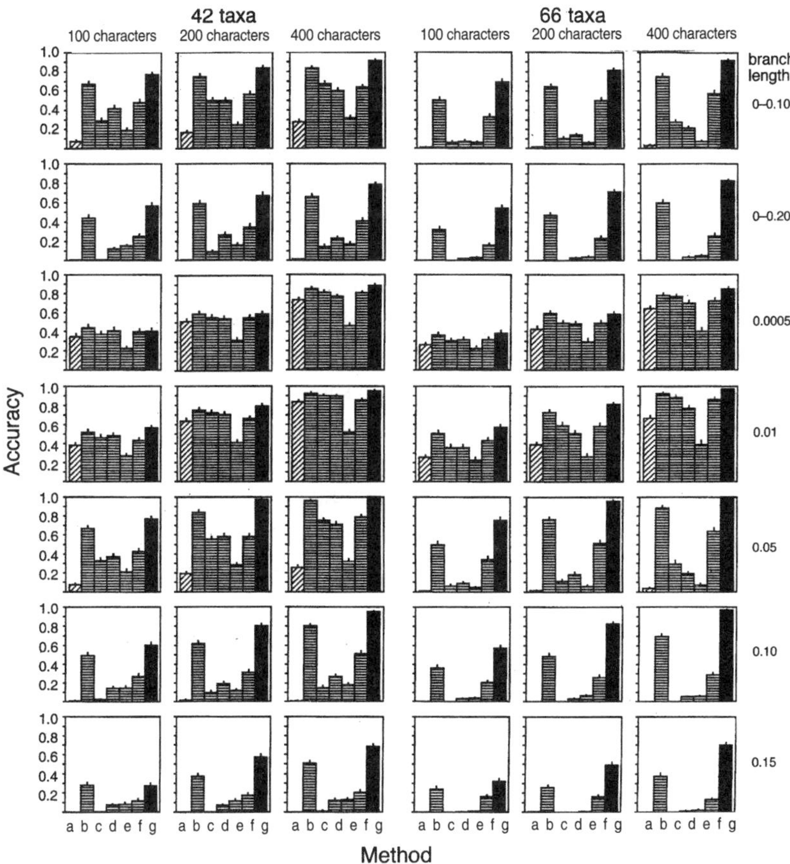

Figure 5.7. Results from simulations showing the accuracy of different methods for coding and sampling interspecific variation in higher taxa. Each bar is the average accuracy from 100 replicated matrixes, and the line extending from the top of the bar represents the standard error. Accuracy is the number of nodes in common between the true and estimated phylogenies. For the simulations represented in the top two rows, branch lengths (probability of change in each character for each lineage) were varied randomly among lineages. Key for methods: a, fixed only; b, majority; c, polymorphic; d, primitive; e, derived; f, single species; g, species as terminals. Figure modified from Wiens (1998a).

tice of coding supraspecific taxa as terminal units. The methods that exclude polymorphic characters (fixed only), render polymorphic data cells uninformative (missing, polymorphic), or arbitrarily fill in data cells with either all zeros or all ones (primitive state, derived state) perform very poorly when many higher taxa are variable and will actually perform worse as more species are sampled (and more variation is detected). These methods do not utilize any information on the distribution of variation within terminals, and they treat all instances of interspecific variation identically. The majority method generally performs better than these methods, seemingly because it uses some information on the distribution of states within the variable higher taxon. In fact, these results and those of further simulations suggest that the common-equals-primitive assumption may have some predictive value (Wiens 1998a). The fixed-only method performs very poorly in the simulations, and its use is made even more questionable by the observation that a real morphological data set with abundant interspecific variation within the terminal taxa (families and other higher taxa of squamate reptiles; Estes et al. 1988) shows significant phylogenetic information in the variable characters and does not generally show a significant relationship between homoplasy and within-taxon variability (Wiens 1998a). The results also show the importance of thorough taxon sampling (even at the expense of sampling more characters), and numerous other studies have likewise emphasized the importance of taxon sampling to phylogenetic accuracy (see Hillis 1998 for a recent review).

Areas for Future Research

The results of recent simulations and previous discussions (Nixon and Davis 1991; Yeates 1995; Bininda-Emonds et al. 1998) strongly support the idea that interspecific variation should be treated by coding species, rather than higher taxa, as terminal units in phylogenetic analyses. However, some authors have recently advocated coding higher taxa as terminal units using inferred ancestral states, or IAS (Donoghue 1994; Mishler 1994; Rice et al. 1997; Bininda-Emonds et al. 1998). The motivation for this approach is to reduce the size of the data matrixes that must be analyzed and facilitate finding the shortest tree. My simulations (Wiens 1998b) did not address the accuracy of the IAS method, because the a priori information used by the IAS method will vary from group to group, and I considered it too difficult to model realistically. Thus, it is unclear whether the advantages of reducing the number of terminal taxa outweigh the loss of information inherent in representing many species as a single taxon. This will likely depend on the number of taxa, number of characters, levels of homoplasy, and other details, and remains an area in need of further study. Nevertheless, my suspicion is that the problems of finding the shortest tree may be minor relative to the loss of information in coding higher taxa as terminal units, especially as computing power and search algorithm speed continue to escalate. I found that heuristic searches with 66 taxa retrieved the correct higher level phylogeny more often than branch-and-bound searches with only six taxa (Wiens 1998a), and Hillis (1996) showed that heuristic searches of simulated data sets with more than 200 taxa could consistently achieve 100% accuracy given a large sample of characters.

Apart from being a hinderance in finding the shortest tree, having large numbers of species included as terminal taxa may also make it more difficult to estimate the reliability of the recovered clades (Rice et al. 1997). The effectiveness of measures of clade support (such as bootstrapping) with large numbers of taxa requires additional study. However, I would argue that large numbers of taxa may simply affect the conservativeness of these measures, rather than invalidating them entirely.

It should be noted that most studies will not include overwhelming numbers of terminal taxa because in most studies hundreds of taxa are not actually sampled. More commonly, a few representative species ("exemplars"—Yeates 1995) are used to represent higher taxa. The simulations of Wiens (1998a) suggest that even when only a few species are sampled from each higher taxon, the species-as-terminals method remains superior to the other coding methods that were tested.

Conclusions and Recommendations

Morphological systematists often use higher level taxa as terminal units in phylogenetic analyses and exclude characters that vary within these units. Simulations (and some empirical data) suggest strongly that breaking up these taxa and using species as terminals will give more accurate results than will coding higher taxa as terminals or excluding characters that vary within them. These two results conflict strongly with much traditional practice in higher level morphological phylogenetics. Nevertheless, they agree closely with studies of the phylogenetic analysis of intraspecifically variable characters. Both show that characters that vary within taxonomic units do contain useful information, and that they should be coded by methods that incorporate as much information as possible about the distribution of this variation.

SUMMARY

Variation within the terminal units of an analysis, whether those units are species or higher level taxa, is ubiquitous in morphological data. Despite this, morphologists rarely describe how this variation is dealt with methodologically in empirical phylogenetic studies (e.g., the criterion used to include or exclude variable characters and how these characters are coded if included). This is an important omission, because different methods for coding this variation can give very different trees for the same data. Furthermore, the simulation and congruence studies reviewed in this chapter suggest that different methods can differ greatly in accuracy (the extent to which the estimated trees resemble the true phylogeny). These analyses suggest that the traditional approach of excluding characters that vary within taxa (the fixed-only method) often performs very poorly, and that there are methods that consistently will perform much better. These include methods that make use of frequency information for intraspecific variation, and the use of species as terminals for interspecific variation. These conclusions also receive some support from a more philosophical perspective, given that excluding characters and frequency information seems to contradict the maxim of total evidence. However, the empirical database on which these conclusions are based is small, mostly because of the scarcity of published data sets that include characters that vary within the terminal units of the analysis. It is hoped that these results will provide an impetus for morphological systematists to routinely include these data and describe their analyses in a more explicit manner.

ACKNOWLEDGMENTS

I thank Tod Reeder and Maria Servedio for collaborating with me on projects related to this paper, and Rich Leschen, Brad Livezey, Paula Mabee, and Maria Servedio for helpful comments and reviews.

REFERENCES

Allard, M. W., and M. M. Miyamoto. 1992. Testing phylogenetic approaches with empirical data, as illustrated with the parsimony method. Molecular Biology and Evolution 9: 778–786.

Berlocher, S. H., and D. L. Swofford. 1997. Searching for phylogenetic trees under the frequency parsimony criterion: An approximation using generalized parsimony. Systematic Biology 46: 211–215.

Bininda-Emonds, O. R. P., H. N. Bryant, and A. P. Russell. 1998. Supraspecific taxa as terminals in cladistic analysis: Implicit assumptions of monophyly and a comparison of methods. Biological Journal of the Linnean Society 64: 101–133.

Brower, A. V. Z., R. deSalle, and A. Vogler. 1996. Gene trees, species trees, and systematics: A cladistic perspective. Annual Review of Ecology and Systematics 27: 423–450.

Campbell, J. A., and D. R. Frost. 1993. Anguid lizards of the genus *Abronia:* Revisionary notes, description of four new species, a phylogenetic analysis, and key. Bulletin of the American Museum of Natural History 216: 1–121.

Carlson, S. J. 1995. Phylogenetic relationships among extant brachiopods. Cladistics 11: 131–197.

Cavalli-Sforza, L. L., and A. W. F. Edwards. 1967. Phylogenetic analysis: Models and estimation procedures. American Journal of Human Genetics 19: 233–257.

Chu, P. C. 1998. A phylogeny of the gulls (Aves: Larinae) inferred from osteological and integumentary characters. Cladistics 14: 1–43.

Coloma, L. A. 1997. Morphology, systematics, and phylogenetic relationships among frogs of the genus *Atelopus* (Anura: Bufonidae). Doctoral dissertation, University of Kansas, Lawrence.

Crother, B. I. 1990. Is "some better than none" or do allele frequencies contain phylogenetically useful information? Cladistics 6: 277–281.

Darwin, C. 1859. On the Origin of Species. Harvard University Press, Cambridge, Mass.

de Queiroz, K. 1987. Phylogenetic systematics of iguanine lizards: A comparative osteological study. University of California Publications in Zoology 118: 1–203.

Domning, D. P. 1994. A phylogenetic analysis of the Sirenia. Proceedings of the San Diego Society of Natural History 29: 177–189.

Donoghue, M. J. 1994. Progress and prospects in reconstructing plant phylogeny. Annals of the Missouri Botanical Gardens 81: 405–418.

Eernisse, D. J., J. S. Albert, and F. E. Anderson. 1992. Annelida and Arthropoda are not sister taxa: A phylogenetic analysis of spiralian metazoan morphology. Systematic Biology 41: 305–330.

Estes, R., K. de Queiroz, and J. A. Gauthier. 1988. Phylogenetic relationships within Squamata. Pages 119–281 *in* Phylogenetic Relationships of the Lizard Families (ed. R. Estes and G. K. Pregill). Stanford University Press, Stanford, Calif.

Farris, J. S. 1966. Estimation of conservatism of characters by constancy within biological populations. Evolution 20: 587–591.

Farris, J. S. 1969. A successive approximations approach to character weighting. Systematic Zoology 18: 374–385.

Farris, J. S. 1981. Distance data in phylogenetic analysis. Pages 3–23 *in* Advances in Cladistics: Proceedings of the First Meeting of the Willi Hennig Society (ed. V. A. Funk and D. R. Brooks). New York Botanical Garden, New York.

Felsenstein, J. 1978. Cases in which parsimony or compatibility methods will be positively misleading. Systematic Zoology 27: 401–410.

Felsenstein, J. 1981. Evolutionary trees from gene frequencies and quantitative characters: Finding maximum likelihood estimates. Evolution 35: 1229–1242.

Felsenstein, J. 1995. PHYLIP, version 3.57c. Department of Genetics, University of Washington, Seattle.

Frost, D. R., and R. Etheridge. 1989. A phylogenetic analysis and taxonomy of iguanian lizards (Reptilia: Squamata). University of Kansas Museum of Natural History Miscellaneous Publications 81: 1–65.

Gaines, M. S., L. R. McClenaghan Jr., and R. K. Rose. 1978. Temporal patterns of allozyme variation in fluctuating populations of *Microtus ochrogaster*. Evolution 32: 723–739.

Gauthier, J., A. G. Kluge, and T. Rowe. 1988. Amniote phylogeny and the importance of fossils. Cladistics 4: 105–209.

Grismer, L. L. 1999. Phylogeny, taxonomy, and biogeography of *Cnemidophorus hyperythrus* and *C. ceralbensis* (Squamata: Teiidae) in Baja California, Mexico. Herpetologica 55: 28–42.

Gutberlet, R. L., Jr. 1998. The phylogenetic position of the Mexican black-tailed pitviper (Squamata: Viperidae: Crotalinae). Herpetologica 54: 184–206.

Harvey, M. B., and R. L. Gutberlet Jr. 1999. A phylogenetic analysis of the Tropidurini (Squamata: Tropiduridae) using new characters of squamation and epidermal microstructure. Zoological Journal of the Linnean Society: in press.

Hillis, D. M. 1991. Discriminating between phylogenetic signal and random noise in DNA sequences. Pages 278–294 *in* Phylogenetic Analysis of DNA Sequences (ed. M. M. Miyamoto and J. Cracraft). Oxford University Press, New York.

Hillis, D. M. 1995. Approaches for assessing phylogenetic accuracy. Systematic Biology 44: 3–16.

Hillis, D. M. 1996. Inferring complex phylogenies. Nature 383: 130–131.

Hillis, D. M. 1998. Taxonomic sampling, phylogenetic accuracy, and investigator bias. Systematic Biology 47: 3–8.

Hollingsworth, B. D. 1998. The systematics of chuckwallas *(Sauromalus)*, with a phylogenetic analysis of other iguanid lizards. Herpetological Monographs 12: 38–191.

Huelsenbeck, J. P. 1995. The performance of phylogenetic methods in simulation. Systematic Biology 44: 17–48.

Kim, J., and M. A. Burgman. 1988. Accuracy of phylogenetic-estimation methods under unequal evolutionary rates. Evolution 42: 596–602.

Kimura, M. 1955. Random genetic drift in multi-allelic locus. Evolution 9: 419–435.

Kluge, A. G. 1989. A concern for evidence and a phylogenetic hypothesis of relationships among *Epicrates* (Boidae, Serpentes). Systematic Zoology 38: 7–25.

Kluge, A. G. 1993. *Aspidites* and the phylogeny of pythonine snakes. Records of the Australian Museum 19: 1–77.

Kluge, A. G. 1997. Testability and the refutation and corroboration of cladistic hypotheses. Cladistics 13: 81–96.

Kluge, A. G., and J. S. Farris. 1969. Quantitative phyletics and the evolution of anurans. Systematic Zoology 18: 1–32.

Kraus, F. 1988. An empirical evaluation of the use of the ontogeny polarization criterion in phylogenetic inference. Systematic Zoology 37: 106–141.

Livezey, B. C. 1986. A phylogenetic analysis of Recent anseriform genera using morphological characters. Auk 103: 737–754.

Livezey, B. C. 1996. A phylogenetic reassessment of the tadornine–anatine divergence (Aves: Anseriformes: Anatidae). Annals of the Carnegie Museum 65: 27–88.

Livezey, B. C. 1997. A phylogenetic analysis of basal Anseriformes, the fossil *Presby-*

ornis, and the interordinal relationships of waterfowl. Zoological Journal of the Linnean Society 121: 361–428.

Mabee, P. M., and J. Humphries. 1993. Coding polymorphic data: Examples from allozymes and ontogeny. Systematic Biology 42: 166–181.

Mayr, E. 1969. Principles of Systematic Zoology. McGraw-Hill, New York.

McGuire, J. A. 1996. Phylogenetic systematics of crotaphytid lizards (Reptilia: Iguania: Crotaphytidae). Bulletin of the Carnegie Museum of Natural History 32: 1–143.

Mickevich, M. F., and M. S. Johnson. 1976. Congruence between morphological and allozyme data in evolutionary inference and character evolution. Systematic Zoology 25: 260–270.

Mickevich, M. F., and C. Mitter. 1981. Treating polymorphic characters in systematics: A phylogenetic treatment of electrophoretic data. Pages 45–58 *in* Advances in Cladistics: Proceedings of the First Meeting of the Willi Hennig Society (ed. V. A. Funk and D. R. Brooks). New York Botanical Garden, New York.

Mickevich, M. F., and C. Mitter. 1983. Evolutionary patterns in allozyme data: A systematic approach. Pages 169–176 *in* Advances in Cladistics, Volume 2: Proceedings of the Second Meeting of the Willi Hennig Society (ed. V. A. Funk and D. R. Brooks). Columbia University Press, New York.

Mink, D. G., and J. W. Sites Jr. 1996. Species-limits, phylogenetic relationships, and origins of viviparity in the *scalaris* complex of the lizard genus *Sceloporus* (Phrynosomatidae: Sauria). Herpetologica 52: 551–571.

Mishler, B. D. 1994. Cladistic analysis of molecular and morphological data. American Journal of Physical Anthropology 94: 143–156.

Miyamoto, M. M., and W. M. Fitch. 1995. Testing species phylogenies and phylogenetic methods with congruence. Systematic Biology 44: 64–76.

Murphy, R. W. 1993. The phylogenetic analysis of allozyme data: Invalidity of coding alleles by presence/absence and recommended procedures. Biochemical Systematics and Ecology 21: 25–38.

Nei, M. 1972. Genetic distance between populations. American Naturalist 106: 238–292.

Nixon, K. C., and J. I. Davis. 1991. Polymorphic taxa, missing values, and cladistic analysis. Cladistics 7: 233–241.

Nixon, K. C., and Q. D. Wheeler. 1990. An amplification of the phylogenetic species concept. Cladistics 6: 211–223.

Novacek, M. J. 1992. Fossils, topologies, missing data, and the higher level phylogeny of eutherian mammals. Systematic Biology 41: 58–73.

Poe, S. 1998. Skull characters and the cladistic relationships of the Hispaniolan dwarf tig *Anolis*. Herpetological Monographs 12: 192–236.

Reeder, T. W., and J. J. Wiens. 1996. Evolution of the lizard family Phrynosomatidae as inferred from diverse types of data. Herpetological Monographs 10: 43–84.

Rice, K. A., M. J. Donoghue, and R. G. Olmstead. 1997. Analyzing large data sets: *rbc*L 500 revisited. Systematic Biology 46: 554–563.

Rodriguez, C. M. 1997. Phylogenetic analysis of the tribe Poeciliini (Cyprodontiformes: Poeciliidae). Copeia 1997: 663–679.

Rowe, T. 1988. Definition, diagnosis, and origin of Mammalia. Journal of Vertebrate Paleontology 8: 241–264.

Schultze, H.-P. 1994. Comparison of hypotheses on the relationships of sarcopterygians. Systematic Biology 43: 155–173.

Shaffer, H. B., P. Meylan, and M. L. McKnight. 1997. Tests of turtle phylogeny: Molecular, morphological, and paleontological approaches. Systematic Biology 46: 235–268.

Siddall, M. E., and A. G. Kluge. 1997. Probabilism and phylogenetic inference. Cladistics 13: 313–336.

Simpson, G. G. 1961. Principles of Animal Taxonomy. Columbia University Press, New York.

Smith, J. F. 1996. Tribal relationships within Gesneriaceae: A cladistic analysis of morphological data. Systematic Botany 21: 497–513.

Stevens, P. F. 1991. Character states, morphological variation, and phylogenetic analysis: A review. Systematic Botany 16: 553–583.

Swofford, D. L. 1993. PAUP (Phylogenetic Analysis Using Parsimony), version 3.1. Illinois Natural History Survey, Champaign.

Swofford, D. L. 1997. PAUP* (Phylogenetic Analysis Using Parsimony*), version 4.0. Sinauer Associates, Sunderland, Mass.

Swofford, D. L., and S. H. Berlocher. 1987. Inferring evolutionary trees from gene frequency data under the principle of maximum parsimony. Systematic Zoology 36: 293–325.

Swofford, D. L., and G. J. Olsen. 1990. Phylogeny reconstruction. Pages 411–501 *in* Molecular Systematics (ed. D. M. Hillis and C. Moritz). Sinauer Associates, Sunderland, Mass.

Swofford, D. L., G. J. Olsen, P. J. Waddell, and D. M. Hillis. 1996. Phylogeny reconstruction. Pages 407–514 *in* Molecular Systematics, 2nd edition (ed. D. M. Hillis, C. Moritz, and B. Mable). Sinauer Associates, Sunderland, Mass.

Thiele, K. 1993. The holy grail of the perfect character: The cladistic treatment of morphometric data. Cladistics 9: 275–304.

Trueb, L., and R. Cloutier. 1991. A phylogenetic investigation of the inter- and intrarelationships of the Lissamphibia (Amphibia: Temnospondyli). Pages 223–313 *in* Origins of the Higher Groups of Tetrapods: Controversy and Consensus (ed. H.-P. Schultze and L. Trueb). Cornell University Press, Ithaca, N.Y.

Vrana, P., and W. Wheeler. 1992. Individuals as terminal entities: Laying the species problem to rest. Cladistics 8: 67–72.

Walker, A. K., I. J. Kitching, and A. D. Austin. 1990. A reassessment of the phylogenetic relationships within the Microgastrinae (Hymenoptera: Braconidae). Cladistics 6: 291–306.

Whiting, M. F., J. C. Carpenter, Q. D. Wheeler, and W. C. Wheeler. 1997. The Strepsiptera problem: Phylogeny of the holometabolous insect orders inferred from 18S and 28S ribosomal DNA sequences and morphology. Systematic Biology 46: 1–67.

Wiens, J. J. 1993a. Phylogenetic systematics of the tree lizards (genus *Urosaurus*). Herpetologica 44: 399–420.

Wiens, J. J. 1993b. Phylogenetic relationships of phrynosomatid lizards and monophyly of the *Sceloporus* group. Copeia 1993: 287–299.

Wiens, J. J. 1995. Polymorphic characters in phylogenetic systematics. Systematic Biology 44: 482–500.

Wiens, J. J. 1998a. The accuracy of methods for coding and sampling higher-level taxa for phylogenetic analysis: A simulation study. Systematic Biology 47: 381–397.

Wiens, J. J. 1998b. Testing phylogenetic methods with tree congruence: Phylogenetic analysis of polymorphic morphological characters in phrynosomatid lizards. Systematic Biology 47: 411–428.

Wiens, J. J., and T. W. Reeder. 1997. Phylogeny of the spiny lizards *(Sceloporus)* based on molecular and morphological evidence. Herpetological Monographs 11: 1–101.

Wiens, J. J., and M. R. Servedio. 1997. Accuracy of phylogenetic analysis including and excluding polymorphic characters. Systematic Biology 46: 332–345.

Wiens, J. J., and M. R. Servedio. 1998. Phylogenetic analysis and intraspecific variation: Performance of parsimony, distance, and likelihood methods. Systematic Biology 47: 228–253.

Yeates, D. K. 1995. Groundplans and exemplars: Paths to the Tree of Life. Cladistics 11: 343–357.

LUCINDA A. MCDADE

6

HYBRIDIZATION AND PHYLOGENETICS

Special Insights from Morphology

Hybridization poses a number of challenges for phylogeneticists. These are perceived by some to be sufficiently serious that it has been argued that groups that are hybrid-prone are not usefully studied using the methods of phylogenetics (e.g., Stebbins 1950; Cronquist 1987). Although relatively few phylogenetic biologists take this extreme stance, it is important to understand these challenges in order to avoid being misled by hybridization and to make progress in dealing with hybrids phylogenetically. In this chapter, after first reviewing the challenges posed by hybrids, I will argue that morphological data and a phylogenetic perspective are central to studies of hybridization in at least four ways. First, morphology plays a key role in the discovery of hybrids: most hybrids have been, and will continue to be, discovered on the basis of morphology. Second, the study of hybrids—specifically, comparisons of the character patterns of hybrids and their parents—offers insight into the most challenging aspect of phylogenetics, character analysis. Third, expectations for the phylogenetic placement of hybrids based on morphology provide a "reality check" on the placement of such taxa based on other character systems and thus a basis for understanding the nature of conflict between phylogenetic studies based on different character systems. Fourth, the evolutionary fate of hybrids is related to their morphological characteristics. For all of these reasons, making progress on the "hybrid problem" demands continued study of hybrids as morphological beings.

THE "HYBRID PROBLEM" IN PHYLOGENETICS

It is important first to establish briefly the nature of the hybrid problem in phylogenetics. (More detailed discussion of each facet of the hybrid problem may be found in the literature cited below.) The methods currently used to reconstruct phylogenetic relationships assume that the evolutionary history of living organisms has been a series of divergent speciation events. (Other algorithms have been proposed or are under development, but none is widely used yet.) On this framework of cladogenesis, character evolution (anagenesis) yields the nested pattern of shared derived characters (synapomorphies) that is the basis of phylogenetic analysis. The problem with hybrids in phylogenetics is that they have reticulating or anastomosing evolutionary histories; their relationships are not correctly depicted by divergent topologies (in figure 6.1, compare trees A and B), nor are these relationships recoverable by nested patterns of synapomorphy. Thus, hybrids will be placed incorrectly by current phylogenetic methods, and their presence may compromise our ability to infer relationships among nonhybrid taxa (Hull 1979; Humphries 1983; Wagner 1983; Wanntorp 1983; Funk 1985).

The placement of hybrids by phylogenetics and the effect that the inclusion of hybrids has on our ability to recover phylogenetic relationships of nonhybrid taxa depend on how the character patterns of hybrid lineages compare with those of their parents and other taxa in the analysis. Depending on the mode of

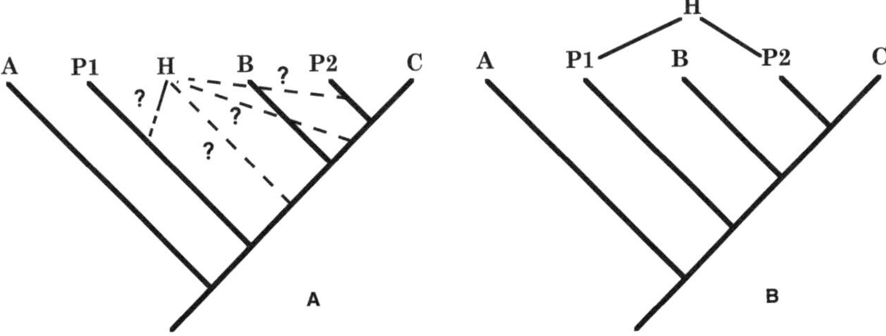

Figure 6.1. Phylogenetic placement of hybrids. **(A)** Phylogenetic analysis will place a hybrid (H) as if it were a divergently evolving taxon; in analyses based on parsimony, its position will be determined by the most parsimonious explanation of its character state patterns relative to those of the other taxa in the analysis. **(B)** The correct depiction of the phylogenetic history of a hybrid is as a reticulation between its parents, P1 and P2.

inheritance and the expression of the characters that are studied, hybrids may have various character patterns (Smith 1992; Rieseberg and Ellstrand 1993; McDade 1995): They may be exactly like one parental taxon (e.g., in uniparentally inherited organellar characters) or may combine the derived characters of both parents (e.g., in biparentally inherited characters with simple expression). Hybrids may also be intermediate between the two parents (in some morphological characters) or heterotic (in some morphological characters).

Each of these patterns of character expression can be used to predict the cladistic placement of hybrids (Smith 1992; McDade 1995). For example, chloroplast or mitochondrial characters should strongly associate a hybrid with the parent from which it received its organelles (assuming unidirectional hybrid origin). Hybrid intermediacy is the dominant pattern in many morphological data sets, and analyses of this type of data will often place a hybrid as the basal member of the lineage of its most derived parent (see McDade [1990] and Smith [1992] for rationale, and McDade [1992] and Eriksson [1995] for examples). This is because the hybrid's intermediate character states will place it between the two parents, whereas the derived states that it shares with the most derived parent (which are likely to outnumber those that it shares with the other, less derived parent) will result in its placement on that parent's lineage.

Regardless of the character system used, however, the phylogenetic methods that are currently employed by systematists cannot place a hybrid correctly (i.e., as a point of reticulation between the parents; figure 6.1B) because such topologies are not among the possible outcomes. Further, there are many reasons that nonhybrid taxa may have character patterns and subsequent phylogenetic "behavior" that are typical of hybrids. For example, ecological convergence may result in a particular taxon combining characters present in two or more other lineages. Likewise, in terms of placement of taxa, any phylogenetic hypothesis with a pectinate or comb-like pattern of relationships will place some taxa in intermediate positions (i.e., between the most basal and most distal branches of the phylogeny), as is expected of hybrids in analyses of morphological characters (McDade 1990, 1992; Smith 1992). Thus, cladistic behavior alone will not identify hybrids or distinguish them unambiguously from taxa with divergent histories.

Knowledge of the special evolutionary history of a hybrid would permit it to be placed correctly post hoc, but such knowledge is rare. More frequently, phylogeneticists must test hypotheses of hybrid origin by using evidence from multiple character systems for which hybrids should behave in predictable and different ways (e.g., Soltis and Soltis 1989; Rieseberg et al. 1990; Novak et al. 1991; Wendel et al. 1991; Dorado et al. 1992). Taxa of hybrid origin may also be discovered on the basis of incongruent results from phylogenetic analysis of data from two or more character systems. For example, it is not uncommon for evidence from chloroplasts or mitochondria to suggest hybridization at some time in the past, even in the absence of supporting evidence from other character systems (e.g., see review by Rieseberg and Soltis 1991; see also Lehman et al. 1991; Wolfe and Elisens 1995; Normark and Lanteri 1998). In other words, data from organelles indicate a pattern of relationships that is incongruent with evidence from morphology and other sources of data. However, there are reasons other than hybrid origin for incongruence between phylogenetic hypotheses based on different character systems (e.g., incomplete lineage sorting of organellar versus nuclear genes). In sum, hybridization should be invoked with care; not all cases of character patterns that are not strictly as expected for divergently evolving taxa are due to hybridization.

As serious as the problem of deducing the correct phylogenetic history of hybrids is the potential impact of hybrids on our ability to unravel relationships among nonhybrid taxa. The presence of hybrid lineages may disrupt the pattern of characters such that relationships among nonhybrid taxa are not correctly recovered or are not resolvable (e.g., Nelson and Platnick 1980; Hill and Crane 1982). For example, strongly discordant patterns of character states in hybrids compared with their parents could act as "noise" in the data set and thereby obscure phylogenetic "signal." Alternatively, the presence of hybrids may result in multiple equally parsimonious hypotheses of phylogenetic relationships involving alternative placements of the hybrid, its parents, and related taxa. When such results are presented as a consensus tree, there may be little resolution of phylogenetic relationships. In my work using morphological characters, I have found the presence of hybrids to be especially problematic for phylogenetic estimation when hybrids are numerous in an analysis (figure 6.2B) or when they are between distantly related taxa (figure 6.2C) (McDade 1992, 1995; see also Bremer and Wanntorp 1979).

In some groups of organisms, hybridization (at least at the level of concern to phylogeneticists: stabilized lineages of hybrid origin) is thought to be sufficiently rare that it can be safely ignored as an evolutionary mode. For example, in his classic works on animal systematics, Mayr (1964, 1970) argued that hybridization is rare in animals and that it does not result in hybrid speciation. In most groups, the empirical basis for this perceived "safety" merits periodic re-examination. Among those who work on higher plants, there is a long history of study of hybrids and concern for their systematic treatment (e.g., Anderson 1936, 1949; Heiser 1949, 1973; Wagner 1968, 1971; Funk 1985). Given that hybrids exist and that we must grapple with them, study of morphological patterns is critical to understanding the incidence and evolutionary role of hybrids, as well as the phylogenetic history of lineages that include hybrids.

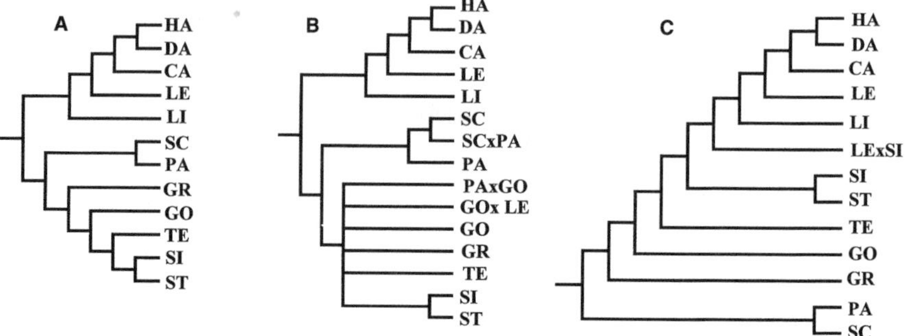

Figure 6.2. Phylogenetic placement of hybrids relative to their parents and related taxa. The example depicted here involves Central American members of the *Aphelandra pulcherrima* complex (Acanthaceae). From McDade (1984). Species abbreviations: CA, *A. campanensis;* DA, *A. darienensis;* GO, *A. golfodulcensis;* GR, *A. gracilis;* HA, *A. hartwegiana;* LE, *A. leonardii;* LI, *A. lingua-bovis;* PA, *A. panamensis;* SC, *A. scabra;* SI, *A. sinclairiana;* ST, *A. storkii;* TE, *A. terryae.* Hybrids are expressed as ovulate parent × staminate parent. **(A)** Most parsimonious hypothesis of the phylogenetic relationships. **(B)** Strict consensus tree of three most parsimonious trees obtained after adding three hybrids. Note that addition of the hybrids causes loss of resolution in the PA × GO through SI + ST clade. **(C)** Single most parsimonious tree from analysis with one added hybrid. The parents of LE × SI are distantly related (see tree A) and the inclusion of this hybrid totally changes the topology (compare trees A and C).

DISCOVERY OF HYBRIDS

The initial discovery of hybridization is usually based on morphology. This is true for the perhaps trivial reason that morphology is readily observed. That is, as little as we know about the morphologies of many organisms, we know a great deal more about their morphologies than about their gene sequences, chromosomes, or enzymes (certain groups of microorganisms that are known exclusively from their gene sequences [e.g., Barns et al. 1996] are obvious exceptions). Hybrids are also likely to be discovered on the basis of morphology for the more important reason that an organism's morphology represents the developmental results of a large sample of the genome. As evidence for hybridization, the overall morphology of an organism is thus more informative than the sequence of one or two genes (see also Doyle 1992). Further, studies that sample from a relatively small portion of the genome may be especially likely to miss hybridization if selection or recombination has occurred. For example, the intriguing results of Rieseberg et al. (1996)—that hybrids between two sunflower species have more genetic material of one parent than the other after only a few generations—suggest that extensive sampling from such a genome might be necessary to detect both parental contributions. In this context, it should be noted that Heiser's (1947) original hypothesis of hybridity for these sunflowers was based on morphology.

In practice, hypotheses regarding hybrid origin of individuals, populations, or taxa usually begin with the observation that the organisms in question are morphologically intermediate between two other taxa. Wagner (1983) has made this point to more than one generation of budding botanists by showing them two of three plants of interest (i.e., two parents plus their hybrid) and sending them into the field to find the third. Botanists, in particular, have used a variety of quantitative approaches to document and display this phenomenon of hybrid intermediacy (e.g., Anderson 1936; Knops and Jensen 1980; Small 1981; Mayer and Mesler 1993).

This prediction of hybrid intermediacy, however, merits careful consideration. In practice, the basis for it is largely that it works. Mechanistically, it works because both parental genetic contributions are integrated in development to produce a hybrid phenotype that is often intermediate. Also at issue is what is meant by intermediacy. We may mean that, for a given character (e.g., corolla length, density of pubescence), hybrids have values that are intermediate between those of the parents. Alternatively, combining the derived parental character states such that hybrids are chimeras has also been described as hybrid intermediacy (e.g., Nelson and Platnick 1980; Funk 1981; Humphries 1983; Humphries and Funk 1984). Mythical hybrids such as centaurs and mermaids are examples of this sort of hybrid intermediacy, whereas mules are examples of the former sort.

My empirical work suggests that, with regard to morphological characters, most hybrid intermediacy is of the former, "mule" type (McDade 1990). Further, my analysis of character expression in hybrids and their parents has indicated that intermediacy on the character level is by far the most frequent pattern in hybrids: "of 103 comparisons among parents and their hybrids in which the hybrid could have been scored as intermediate for a given character (i.e., the parental scores differed by two or more states), hybrids were in fact intermediate in 76 instances or about 75% of the cases" (McDade 1990). Thus, intermediacy dominates character expression in hybrids, and it is in this sense that I use the term "hybrid intermediacy" to refer to the morphology of hybrids throughout the rest of this chapter.

Beyond the issues regarding the nature of hybrid intermediacy, it is clear that looking for hybrids on the basis of intermediacy is somewhat circular. Almost any group of taxa, even if all members are divergently evolving, will include some taxa that are morphologically extreme and others that are intermediate (e.g., almost any phylogenetic analysis of plants will include some species with relatively large flowers, others with smaller flowers, and still others with flowers that are intermediate between the extremes in the group under study). Thus, hypotheses regarding hybridity that are based on morphology require additional testing. Traditional approaches to obtaining further evidence include studies of pollen fertility and chromosomes, as well as breeding experiments (e.g., attempting to synthesize hybrids, selfing or back-crossing putative hybrids to produce future generations). More recently, support for hypotheses of hybrid origin has also come from genetic sources (e.g., examining nuclear characters for markers unique to each putative parent or organellar DNA characters for markers unique to one parent).

There can be little doubt that, as useful as morphology is in the discovery of hybrids, there are instances of hybridization that are cryptic from this perspective. For example, extensive studies of hybrid zones indicate that morphology often provides an incomplete picture of gene exchange (i.e., hybrid zones tend to be wider for genetic than for morphological characters; Patton et al. 1979; Barton and Hewitt 1985, 1989; Paige et al. 1991; Patton 1993). Numerous cases are now known in which organellar data suggest reproductive interactions in the past between taxa for which there is no other evidence of hybridization (e.g., "chloroplast capture," Rieseberg and Soltis 1991). This discovery is consistent with the clonal inheritance of organelles and suggests that selection may sometimes favor heterospecific organelles over conspecific ones. Multi-faceted approaches will be essential to estimate the incidence and significance of this sort of cryptic hybridization.

Derived hybrids (i.e., hybrid lineages that have accumulated apomorphies and may have speciated since their origin) may also be cryptic from the point of view of morphology. These lineages are likely to accumulate apomorphies that eventually will mask any evidence of reticulate evolutionary history. It may in fact be harder to detect ancient and highly derived lineages of hybrid origin on the basis of morphology than by other character systems. Identification of such higher level hybrid lineages may require slowly evolving characters that will maintain the signal of hybridization long after the event has occurred. In allopolyploid groups, chromosomes often provide such a signal. For example, the hybrid origin of catostomid fishes was first proposed because catostomids have twice the number of chromosomes and amount of DNA of related cypriniform fishes (Uyeno and Smith 1972; Ferris 1984). Morphologically, these fishes are highly autapomorphic, and it is unlikely that researchers would have been able to detect their hybrid origin on this basis. Among plants, the hypothesis of the hybrid origin of maloid roses is supported by chromosomes, chemistry, and allozymes (reviewed by Phipps et al. 1991). Here too, morphology is either highly autapomorphic (e.g., fruit type, ovary position) or is shared by other groups of roses (e.g., leaves, growth form).

It is remarkable that, despite considerable evidence for hybrid origin at the species level in plants, very few hypotheses of hybrid origin of species-rich lineages are supported by more than one source of evidence. (The case of the maloid roses just described is perhaps the most strongly supported hypothesis of this sort among plants.) This may well be because morphology is usually our first tool for detecting hybrids. Lineages of hybrid origin may be expected to have autapomorphic morphologies as a result of having evolved independently from the parental lineages after hybridization, and thus to elude recognition as hybrids on the basis of morphology alone. This seems to be the case with catastomid fishes and maloid roses. Further, especially among plants, clear cases of polyploidy are taken to indicate that hybrid origin is likely (e.g., Kellogg and Watson 1993). However, euploid change in chromosome number is a rampant feature of plant evolution and ploidy can vary wildly at low taxonomic levels. Many plant genera include species with a wide range of ploidy levels (see Darlington and Wylie 1955; Moore 1982) and ploidy levels can even vary within taxonomic species (e.g., *Atriplex confertifolia* varies at least from $2n$ to $10n$ [Stutz and Sanderson 1983]). Assuming that most of these polyploids are allopolyploids, these patterns certainly suggest that hybridization is a mode of speciation in plants. However, the baroque nature of chromosomal evolution at low taxonomic levels may mask higher level patterns that would permit detection of lineages of hybrid origin. That is, in many plant groups, chromosome numbers are too evolutionarily vagile to provide the stable signal needed to detect ancient hybridization.

HYBRID MORPHOLOGIES YIELD INSIGHT INTO CHARACTER ANALYSIS

One of the most difficult and also most interesting aspects of phylogenetic analysis based on morphology is character analysis (i.e., assessment of homologies, partitioning of character states, decisions regarding ordering). When we treat variation in a morphological feature as states of a single character, we are hypothesizing that these represent variation in a homologous character (i.e., that they are the result of descent, with modification, of the same genetic material) and thus contain phylogenetic information.

By definition, hybrids combine the genetic material of both parents. The hybrid phenotype that develops from that genetic material is potentially informa-

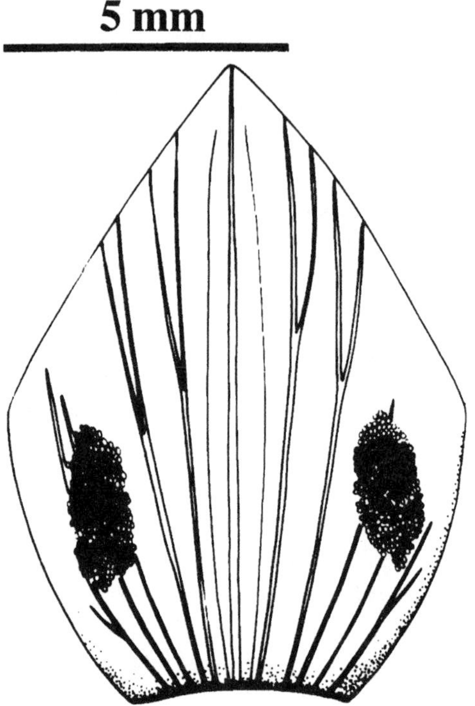

Figure 6.3. Abaxial surface of cleared and stained floral bract of *Aphelandra lingua-bovis* showing paired bracteal nectaries. The nectaries illustrated here are composed of the "small gland" type (see figure 6.4). The position and general appearance of nectaries are the same in all *Aphelandra* species that have them, regardless of their gland type.

tive regarding hypotheses on homology. When interaction between the parental genomes yields an interpretable intermediate character state or yields only one or the other parental phenotype, the result is consistent with a hypothesis of homology. The hybrid may also express both parental characters, which I take to be an uninformative outcome from the perspective of homology decisions. In contrast, Patterson (1982) has argued that such characters are not homologous because this outcome fails his test of conjunction (i.e., co-occurrence in one organism). However, if a trait has a relatively simple genetic basis, the parental conditions could be codominant, such that both might be apparent in the hybrid.

Plants of the genus *Aphelandra* (Acanthaceae) serve as an example of how hybrid phenotypes can inform hypotheses on homology. Members of a monophyletic lineage within this genus have nectaries on the floral bracts (Leonard 1953; McDade and Turner 1997; figure 6.3). As documented by McDade and Turner (1997), these nectaries comprise individual glands that vary in size (68–783 μm in diameter), in number per patch (one to several hundred glands) and, most notably, in number of cells in the intermediate layer of the gland's three cell layers. Members of two distinct sublineages have either a single cell or many cells in the intermediate layer of the gland (figure 6.4); this difference is accompanied by differences in size of individual glands and in number of glands per nectarial patch (figure 6.4; table 6.1). Four hybrids were synthesized by crossing members of these two lineages (McDade 1984, 1992), and they were examined for nectary morphology and anatomy with the methods described by McDade and Turner (1997). The glands of these hybrid plants are intermediate in size, number per patch, and cellularity of the intermediate layer (table 6.1; figure 6.4). I take these data as support for the hypothesis that the two kinds of parental nectaries are homologous: the parental genomes interact to yield an intermediate phenotype in the hybrid.

Hybrid phenotypes also provide insight into decisions regarding ordering of multistate characters. When such characters are treated as ordered (or "maximally connected" in the terminology of Slowinski [1993]), more evolutionary change is required between the extreme character states than from an intermediate state to either extreme. (Note that branched ordering schemes are also possible—see Slowinski 1993.) In general, ordering increases the power of phylogenetic analyses to resolve relationships (Slowinski 1993; but see Hauser and Presch 1991) and thus is advantageous, assuming that resolved phylogenetic hypotheses are desirable and that we have some defensible rationale for ordering.

Hauser and Presch (1991) present a very useful overview of the rationales used by systematists to order the states of multistate characters. These authors indicate that similarity is often the basis for ordering; that is, the condition of a particular character that is judged to represent the intermediate state is more similar to each of the extreme character states than the extremes are to each other. I would extend these ideas to suggest that ordering multistate characters based on similarity represents an assumption about the genetic basis of the character in question: Phenetic intermediates between character extremes are hypothesized to be genetic intermediates. Hybrids are, by definition, genetic intermediates between the parents, and their characters therefore have an intermediate genetic basis. If a hybrid's intermediate morphology for a particular character matches that of other taxa included in a phylogenetic analysis, evidence from the hybrid supports ordering the states of that character. Thus, in the example from *Aphelandra* described above, compared with their parents, hybrids are intermediate in number of glands, size of glands, and number of cells

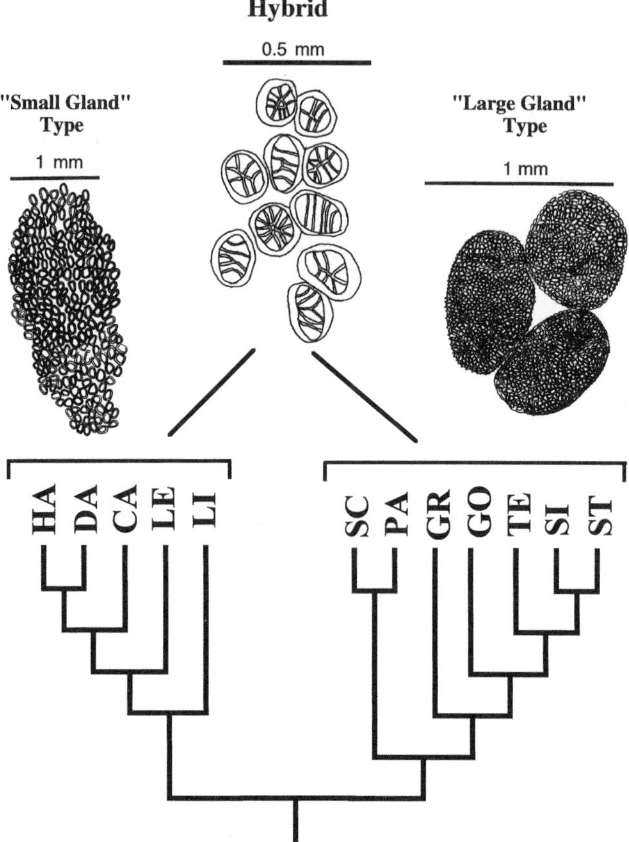

Figure 6.4. Extrafloral nectary gland morphologies of the two main lineages of the *Aphelandra pulcherrima* complex (Acanthaceae) and their hybrids. The thick walls of the cell(s) of the intermediate layer of the glands are illustrated ("small gland" type drawn from *A. lingua-bovis;* "large gland" type drawn from *A. scabra;* hybrid drawn from *A. leonardii* × *A. sinclairiana*). Hybrids with one parent from each lineage have glands that are intermediate in size, in number per nectary, and in number of cells in the intermediate layer (see table 6.1).

in the intermediate layer. These data support ordering the states of these three characters in the data set devised for phylogenetic analysis of *Aphelandra* species.

The phylogenetic use of hybrid morphological characters builds on the more general fact that, as has been known by plant and animal breeders at least since

Table 6.1
Glands of extrafloral nectaries of hybrids and their parents

	No. glands per nectary			Gland diameter (μm)		No. cells in intermediate layer[b]
Taxa[a]	Mean	Range	*n*	Mean ± SD	*n*	
"Large gland" parents						
PA	2.2	1–5	12	450 ± 77	20	>100
GO	4.3	2–6	11	520 ± 94	12	>100
SI	3.9	2–7	17	488 ± 77	21	>100
"Small gland" parent, LE	68.4	55–77	5	148 ± 22	9	1
Hybrids						
GO × LE	10.7	8–14	9	258 ± 42	23	22
LE × GO	14.3	7–25	18	199 ± 18	20	12
LE × SI, plant 1	9.0	3–18	18	276 ± 40	20	18
LE × SI, plant 2	20.6	13–43	9	206 ± 26	27	10
PA × LE	41.1	29–52	9	235 ± 31	20	?

Parents are species of *Aphelandra* (Acanthaceae) that differ in gland type. Full names of the species are listed in the caption to figure 6.2, and the three gland morphs are illustrated in figure 6.4.

[a]See McDade and Turner (1997) for voucher data.

[b]For hybrids, values are means.

Mendel, hybrids provide insight into the genetic basis of morphological traits. Exploration of the genetic basis of traits by applying the methods of quantitative trait locus (QTL) mapping to the characters of hybrids between distinct parents is a growth industry. The QTL approach has been extremely useful, for example, in exploring the genetic basis of differences in floral morphology (Bradshaw et al. 1995) and the genetic basis of morphological characters that distinguish maize from its closest relatives (Doebley and Stec 1993; Dorweiler et al. 1993; Doebley et al. 1995). These techniques have also enabled examination of the genetic basis for the heterosis that is observed in some characters of some hybrids (reviewed by Lynch and Walsh 1997).

MORPHOLOGICAL PATTERNS PREDICT PHYLOGENETIC PLACEMENT

The phenotypes of taxa, as understood by systematists working with a particular group of organisms, represent predictions regarding their phylogenetic placement. These predictions are rarely formalized: phylogeneticists do not devise elaborate predictions about, for example, the placement of humans with other great apes or of sunflowers with other composites. However, the placement of humans with rodents or of sunflowers with pines would certainly signal a problem. Thus, a species or lineage that is morphologically intermediate in the context of the group being studied is expected to be placed in a phylogenetically intermediate position. Further, such a lineage should be placed in this manner regardless of the character system (or systems) used. That is, barring various processes that yield incongruence among character sets (e.g., selection on some traits but not others, incomplete lineage sorting, hybridization), and assuming one has sufficient data, phylogeneticists should recover the same pattern of divergent evolution regardless of the character system used.

If a lineage that is morphologically intermediate is placed in a radically different position in an analysis based on another character system, the conflict of findings signifies that further investigation is needed. For example, chloroplast or mitochondrial DNA characters would place a morphologically intermediate taxon of hybrid origin as the sister taxon to its (usually) maternal parent. Such an outcome would certainly guide the careful researcher to consider the evolutionary history of this taxon. In this light, Rieseberg and Soltis (1991) have warned against the blind use of organellar characters in phylogeny reconstruction. This also provides a very strong argument for knowing one's organisms morphologically regardless of which character systems one's research emphasizes.

It should be noted that this warning holds whether or not hybridization is involved. Other processes can yield data that mislead. For example, under a variety of circumstances, gene trees may not be congruent with species trees (e.g., because of incomplete lineage sorting, gene duplication, or hybridization) or they may simply be inaccurate (e.g., because of inadequate sampling, slow rates of evolution, shallow divergence times). Again, phylogenetic outcomes that contradict morphological evidence will often be the basis for recognizing that something is amiss.

A number of efforts are underway to devise algorithms that search for, or at least permit, reticulation (Alroy 1995; Rieseberg and Morefield 1995; A. Dickerman, pers. comm.). Methods that consider the possibility of recombination in genetic data (e.g., Hein 1990, 1993; Templeton et al. 1992) are perhaps adaptable to phylogenetic problems involving hybrids as well. Like other tree-building algorithms, these work from a particular model of character distribution that is expected to mark evolutionary history. The algorithms referred to above that deal explicitly with hybrids look for reticulation by seeking taxa that share the derived character states of two other taxa present in the analysis, such that connecting one taxon to two others improves the fit of the data to the evolutionary hypothesis. This is certainly a useful pattern to look for, but it is important to note that only for certain character systems are hybrids expected to combine the derived states of both parents (e.g., biparentally inherited characters with a simple basis: DNA, chromosomes, allozymes). This sort of character pattern is not likely in morphological data. In fact, it may be especially difficult to modify phylogenetic algorithms to look for hybridization based on morphological characters unless we can differentiate between hybrid intermediacy and the intermediacy expected in any data set with more than a very small number of taxa.

MORPHOLOGY IS A KEY DETERMINANT OF THE FUTURE OF HYBRID LINEAGES

The evolutionary significance of hybridization depends upon the extent to which hybrids can survive, compete, find suitable mates, and reproduce (or last long enough via sheer longevity or asexual reproduction for mutations to restore fertility). Clearly, hybrids that are unable to survive and compete in nature as morphological beings have no evolutionary future. Just as morphology is part of the key to success or failure of hybrids in nature, having useful or aesthetically appealing morphological traits is also key to the success of hybrids made by human intervention. Hybrid corn is grown for its morphological characters; although these are of course genetically based, the corn is not grown because it has interesting genes. (Various laboratory strains of model organisms are exceptions in that they are maintained because of their genes.) Heterosis is reasonably common in hybrids (Rieseberg and Ellstrand 1993, and references therein) and often is responsible for the traits that make a hybrid lineage attractive to humans. It hardly needs to be said that such heterosis is a statement about morphology.

As the exceptions that prove the rule, hybrids that have valuable traits but are not especially robust or competitive in nature require our protection and would certainly go extinct without it (e.g., hybrid roses, many crop plants). Similarly, humans regularly create hybrids that cannot reproduce (e.g., bananas that are triploid; corn that cannot disperse its seeds without intervention). The cost of availing ourselves of the useful morphological characters of these hybrids is that they must be propagated vegetatively or resynthesized at each generation.

CONCLUSION

Although data from macromolecules will continue to contribute a great deal to the study of hybrids, morphology must be kept ever in mind. Morphology is key to the recognition of most hybrids and provides a "reality check" on the expected placement of hybrids in phylogenetic analysis. This verification may be important, especially in hybrid-prone lineages, if we are to avoid avoidable errors in phylogenetic estimation. Hybrids are also extremely useful to phylogeneticists for the insight that they provide into the dicey issues of character analysis, especially homology and ordering of multistate characters. Finally, there can be little doubt that the overall evolutionary significance of hybridization has been mediated by morphology in the sense that morphologically disastrous and unfit organisms would not have survived to trouble us by demanding that we consider reticulate evolution a facet of life on Earth.

ACKNOWLEDGMENTS

I thank J. Wiens for convening the symposium at which these ideas were presented and for editing this volume, and the Society for Systematic Biology for its support of the symposium. R. Calvert, A. Cardoni, A. Chetochine, and M. D. Turner helped in preparation of the figures, and V. Funk, R. Levin, J. G. Lundberg, J. Miller, G. Smith, and J. Wiens provided valuable comments on earlier versions of the manuscript.

REFERENCES

Alroy, J. 1995. Continuous track analysis: A new phylogenetic and biogeographic method. Systematic Biology 44: 152–178.

Anderson, E. 1936. Hybridization in American Tradescantias. Annals of the Missouri Botanical Garden 23: 511–525.

Anderson, E. 1949. Introgressive Hybridization. Wiley, New York.

Barns, S. M., C. F. Delwiche, J. D. Palmer, and N. R. Pace. 1996. Perspectives on archeal diversity, thermophily, and monophyly from environmental rRNA sequences. Proceedings of the National Academy of Science, U.S.A. 93: 9188–9193.

Barton, N. H., and G. M. Hewitt. 1985. Analysis of hybrid zones. Annual Review of Ecology and Systematics 16: 113–148.

Barton, N. H., and G. M. Hewitt. 1989. Adaptation, speciation, and hybrid zones. Nature 341: 497–503.

Bradshaw, H. D., Jr., S. M. Wilbert, K. G. Otto, and D. W. Schemske. 1995. Genetic mapping of floral traits associated with reproductive isolation in monkeyflowers *(Mimulus).* Nature 376: 762–765.

Bremer, K., and H.-E. Wanntorp. 1979. Hierarchy and reticulations in systematics. Systematic Zoology 28: 624–627.

Cronquist, A. 1987. A botanical critique of cladism. Botanical Review 53: 1–52.

Darlington, C. D., and A. P. Wylie. 1955. Chromosome Atlas of Flowering Plants. George Allen and Unwin, London.

Doebley, J., and A. Stec. 1993. Inheritance of the morphological differences between maize and teosinte: Comparison of results for two F2 populations. Genetics 134: 559–570.

Doebley, J., A. Stec, and B. Kent. 1995. Suppression of sessile spikelets (*Sos*1): A dominant mutant affecting inflorescence development in maize. American Journal of Botany 82: 571–577.

Dorado, O., L. H. Rieseberg, and D. M. Arias. 1992. Chloroplast DNA introgression in southern California sunflowers. Evolution 46: 566–572.

Dorweiler, J., A. Stec, J. Kermicle, and J. Doebley. 1993. Teosinte glume architecture. 1. A genetic locus controlling a key step in maize evolution. Science 262: 233–235.

Doyle, J. J. 1992. Gene trees and species trees: Molecular systematics as one-character taxonomy. Systematic Botany 17: 144–163.

Eriksson, T. 1995. The genus *Athroisma* (Asteraceae, Heliantheae). Botanical Journal of the Linnean Society 119: 101–184.

Ferris, S. D. 1984. Tetraploidy and the evolution of the catostomid fishes. Pages 55–93 *in* Evolutionary Genetics of Fishes (ed. B. J. Turner). Plenum Press, New York.

Funk, V. A. 1981. Special concerns in estimating plant phylogenies. Pages 73–86 *in* Advances in Cladistics: Proceedings of the First Meeting of the Willi Hennig Society (ed. V. A. Funk and D. R. Brooks). New York Botanical Garden, New York.

Funk, V. A. 1985. Phylogenetic patterns and hybridization. Annals of the Missouri Botanical Garden 72: 681–715.

Hauser, D. L., and W. Presch. 1991. The effect of ordered characters on phylogenetic reconstruction. Cladistics 7: 243–265.

Hein, J. 1990. Reconstructing evolution of sequences subject to recombination using parsimony. Mathematical Biosciences 98: 185–200.

Hein, J. 1993. A heuristic method to reconstruct the history of sequences subject to recombination. Journal of Molecular Evolution 36: 396–405.

Heiser, C. B., Jr. 1947. Hybridization between the sunflower species *Helianthus annuus* and *H. petiolaris.* Evolution 1: 249–262.

Heiser, C. B., Jr. 1949. Natural hybridization with particular reference to introgression. Botanical Review (Lancaster) 15: 645–687.

Heiser, C. B., Jr. 1973. Introgression re-examined. Botanical Review (Lancaster) 39: 347–366.

Hill, C. R., and P. R. Crane. 1982. Evolutionary cladistics and the origin of angiosperms. Pages 269–361 *in* Problems of Phylogenetic Reconstruction (ed. K. A. Joysey and A. E. Friday). Academic Press, London.

Hull, D. L. 1979. The limits of cladism. Systematic Zoology 28: 416–440.

Humphries, C. J. 1983. Primary data in hybrid analysis. Pages 89–103 *in* Advances in Cladistics: Proceedings of the Second Meeting of the Willi Hennig Society (ed. N. I. Platnick and V. A. Funk). Columbia University Press, New York.

Humphries, C. J., and V. A. Funk. 1984. Cladistic methodology. Pages 323–361 *in* Current Topics in Plant Taxonomy (ed. V. H. Heywood and D. M. Moore). Academic Press, London.

Kellogg, E. A., and L. Watson. 1993. Phylogenetic studies of a large data set. I. Bambusoideae, Andropogonodae, and Pooideae (Gramineae). Botanical Review 59: 273–343.

Knops, J. F., and R. J. Jensen. 1980. Morphological and phenolic variation in a three species community of red oaks. Bulletin of the Torrey Botanical Club 107: 418–428.

Lehman, N., A. Eisenhawer, K. Hansen, L. D. Mech, R. O. Peterson, P. J. P. Gogan, and R. K. Wayne. 1991. Introgression of coyote mitochondrial DNA into sympatric North American gray wolf populations. Evolution 45: 104–119.

Leonard, E. C. 1953. The Acanthaceae of Colombia, II. Contributions from the U.S. National Herbarium 31: 119–322.

Lynch, M., and B. Walsh. 1997. Genetics and Analysis of Quantitative Characters. Sinauer Associates, Sunderland, Mass.

Mayer, M. S., and M. R. Mesler. 1993. Morphometric evidence of hybrid swarms in mixed populations of *Polystichum munitum* and *P. imbricans* (Dryopteridaceae). Systematic Botany 18: 248–260.

Mayr, E. 1964. Systematics and the Origin of Species from the Viewpoint of a Zoologist. Dover Publications, New York.

Mayr, E. 1970. Populations, Species, and Evolution. Belknap Press of Harvard University Press, Cambridge, Mass.

McDade, L. A. 1984. Systematics and reproductive biology of the Central American species of the *Aphelandra pulcherrima* complex (Acanthaceae). Annals of the Missouri Botanical Garden 71: 104–165.

McDade, L. A. 1990. Hybrids and phylogenetic systematics. I. Patterns of character expression in hybrids and their implications for cladistic analysis. Evolution 44: 1685–1700.

McDade, L. A. 1992. Hybrids and phylogenetic systematics. II. The impact of hybrids on cladistic analysis. Evolution 46: 1329–1346.

McDade, L. A. 1995. Hybridization and phylogenetics. Pages 305–331 *in* Experimental and Molecular Approaches to Plant Biosystematics (ed. P. C. Hoch and A. G. Stephenson). Missouri Botanical Garden Press, Saint Louis.

McDade, L. A., and M. D. Turner. 1997. Structure and development of bracteal nectary glands in *Aphelandra* (Acanthaceae). American Journal of Botany 81: 1–15.

Moore, D. M. 1982. Flora Europaea Check-List and Chromosome Index. Cambridge University Press, Cambridge.

Nelson, G., and N. I. Platnick. 1980. Multiple branching in cladograms: Two interpretations. Systematic Zoology 29: 86–91.

Normark, B. B., and A. A. Lanteri. 1998. Incongruence between morphological and mitochondrial DNA characters suggests hybrid origins of parthenogenetic weevil lineages (genus *Aramigus*). Systematic Biology 47: 475–494.

Novak, S. J., D. E. Soltis, and P. S. Soltis. 1991. Ownby's *Tragopogons:* 40 years later. American Journal of Botany 78: 1586–1600.

Paige, K. N., W. C. Capman, and P. Jenetten. 1991. Mitochondrial inheritance patterns across a cottonwood hybrid zone: Cytonuclear disequilibria and hybrid zone dynamics. Evolution 45: 1360–1369.

Patterson, C. 1982. Morphological characters and homology. Pages 221–274 *in* Problems of Phylogenetic Reconstruction (ed. K. A. Joysey and A. E. Friday). Academic Press, London.

Patton, J. L. 1993. Hybridization and hybrid zones in pocket gophers (Rodentia, Geomyidae). Pages 290–308 *in* Hybrid Zones and the Evolutionary Process (ed. R. G. Harrison). Oxford University Press, New York.

Patton, J. L., J. C. Hafner, M. S. Hafner, and M. F. Smith. 1979. Hybrid zones in *Thomomys bottae* pocket gophers: Genetic, phenetic, and ecologic concordance patterns. Evolution 33: 860–876.

Phipps, J. B., K. R. Robertson, J. R. Rohrer, and P. G. Smith. 1991. Origins and evolution of subfam. Maloideae (Rosaceae). Systematic Botany 16: 303–332.

Rieseberg, L. H., and N. C. Ellstrand. 1993. What can molecular and morphological markers tell us about plant hybridization? Critical Reviews in Plant Science 12: 213–241.

Rieseberg, L. H., and J. D. Morefield. 1995. Character expression, phylogenetic reconstruction, and the detection of reticulate evolution. Pages 333–353 *in* Experimental and Molecular Approaches to Plant Biosystematics (ed. P. C. Hoch and A. G. Stephenson). Missouri Botanical Garden, Saint Louis.

Rieseberg, L. H., and D. E. Soltis. 1991. Phylogenetic consequences of cytoplasmic gene flow in plants. Evolutionary Trends in Plants 5: 65–84.

Rieseberg, L. H., R. Carter, and S. Zona. 1990. Molecular tests of the hypothesized hybrid origin of two diploid *Helianthus* species (Asteraceae). Evolution 44: 1498–1511.

Rieseberg, L. H., B. Sinervo, C. R. Linder, M. C. Ungerer, and D. M. Arias. 1996. Role of gene interactions in hybrid speciation: Evidence from ancient and experimental hybrids. Science 272: 741–745.

Slowinski, J. B. 1993. "Unordered" versus "ordered" characters. Systematic Biology 42: 155–165.

Small, E. 1981. A numerical analysis of morpho-geographic groups of cultivars of *Humulus lupulus* based on samples of cones. Canadian Journal of Botany 59: 311–324.

Smith, G. R. 1992. Introgression in fishes: Significance for paleontology, cladistics, and evolutionary rates. Systematic Biology 41: 41–57.

Soltis, D. E., and P. S. Soltis. 1989. Allopolyploid speciation in *Tragopogon:* Insights from chloroplast DNA. American Journal of Botany 76: 1119–1124.

Stebbins, G. L. 1950. Variation and Evolution in Plants. Columbia University Press, New York.

Stutz, H. E., and S. C. Sanderson. 1983. Evolutionary studies of *Atriplex:* Chromosome races of *A. confertifolia* (shadscale). American Journal of Botany 70: 1536–1547.

Templeton, A. R., K. A. Crandall, and C. F. Sing. 1992. A cladistic analysis of phenotypic associations with haplotypes inferred from restriction endonuclease mapping and DNA sequence data. III. Cladogram estimation. Genetics 132: 619–633.

Uyeno, T., and G. R. Smith. 1972. Tetraploid origin of the karyotype of catostomid fishes. Science 175: 644–646.

Wagner, W. H., Jr. 1968. Hybridization, taxonomy, and evolution. Pages 113–138 *in* Modern Methods in Plant Taxonomy (ed. V. H. Heywood). Academic Press, London.

Wagner, W. H., Jr. 1971. Evolution of *Dryopteris* in relation to the Appalachians. Pages 147–192 *in* The Distributional History of the Southern Appalachians, Part II: Flora (ed. P. C. Holt). Monograph 2 of the Research Division, Virginia Polytechnical Institute State University, Blacksburg.

Wagner, W. H., Jr. 1983. Reticulistics: The recognition of hybrids and their role in cladistics and classification. Pages 63–79 *in* Advances in Cladistics: Proceedings of the Second Meeting of the Willi Hennig Society (ed. N. I. Platnick and V. A. Funk). Columbia University Press, New York.

Wanntorp, H.-E. 1983. Reticulated cladograms and the identification of hybrid taxa. Pages 81–88 *in* Advances in Cladistics: Proceedings of the Second Meeting of the Willi Hennig Society (ed. N. I. Platnick and V. A. Funk). Columbia University Press, New York.

Wendel, J. F., J. M. Steward, and J. H. Rettig. 1991. Molecular evidence for homoploid reticulate evolution among Australian species of *Gossypium.* Evolution 45: 694–711.

Wolfe, A. D., and W. J. Elisens. 1995. Evidence of chloroplast capture and pollen-mediated gene flow in *Penstemon* sect. *Peltanthera* (Scrophulariaceae). Systematic Botany 20: 395–412.

JOHN P. HUELSENBECK AND BRUCE RANNALA

7

USING STRATIGRAPHIC INFORMATION IN PHYLOGENETICS

Perhaps the greatest single advance in systematics over the past 50 years has been the introduction of explicit methods for reconstructing phylogeny. Before about 1950, the construction of phylogenetic trees was entrusted only to those individuals with an expert knowledge of specific taxonomic groups. These experts were often paleontologists who constructed evolutionary trees such as the one shown in figure 7.1. Early phylogenies were constructed mostly on the basis of experts' intuitions about relationships among groups. Although today few systematists would argue that phylogenies should be reconstructed solely on intuition, it is interesting to note that early systematists frequently incorporated additional sources of information into their phylogenetic descriptions that modern systematists now rarely consider; the specific positions of fossil species in the stratigraphic record, for example, were usually taken into account. Early phylogenetic trees, such as the phylogeny of the elephants shown in figure 7.1, were generally consistent with the temporal distributions of fossil species.

The temporal order of fossil species is rarely considered in contemporary phylogenetic analyses that are based on DNA sequences or morphology of extant species. In fact, only a handful of modern researchers have either directly incorporated stratigraphic data into a phylogenetic analysis or investigated the agreement of phylogenetic trees with the stratigraphic record (Harper 1976; Gingerich 1979; Fisher 1982, 1988, 1991, 1992, 1994; Gauthier et al. 1988; Marshall 1990; Norell and Novacek 1992a, 1992b; Huelsenbeck 1994; Benton 1995; Wagner 1995; Benton and Storrs 1996; Brochu 1997; Huelsenbeck and Rannala 1997). Moreover, explicit methods for incorporating stratigraphic information into a phylogenetic analysis have been developed only recently, and

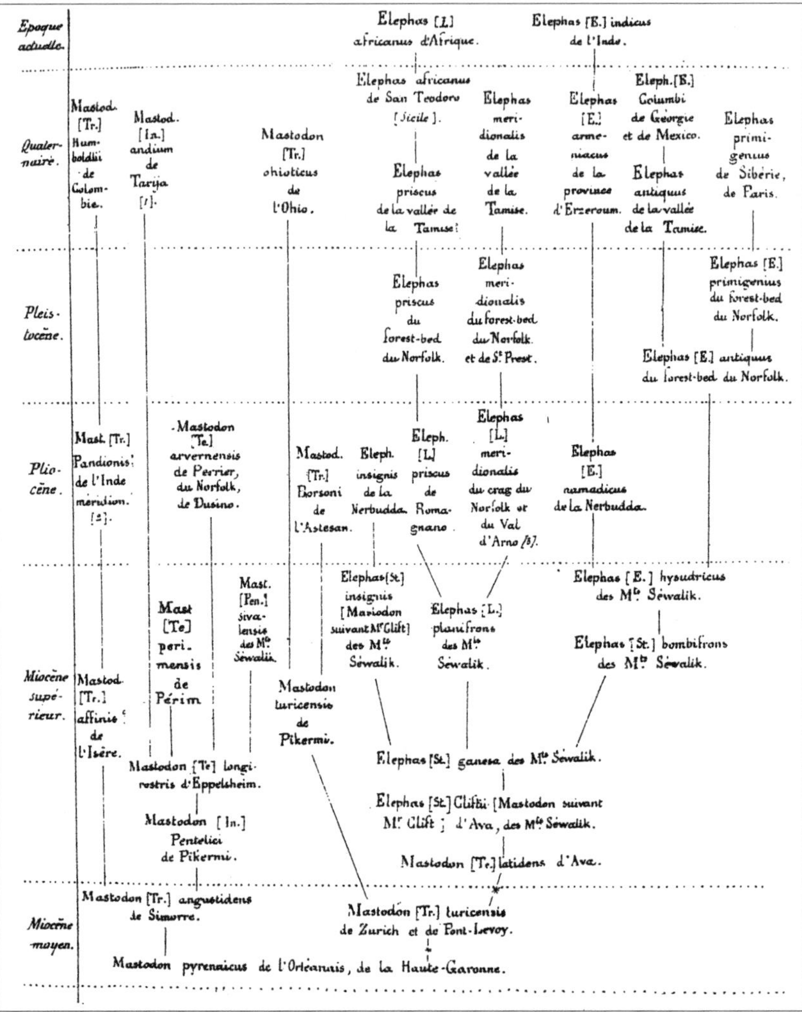

Figure 7.1. A phylogeny of elephants proposed by Gaudry (1862) based, in large part, on collections of Miocene mammals from Pikermi, Greece. Gaudry's phylogenies of mammals are among the first published after the *Origin of Species* (Darwin 1859). Note that the phylogenetic tree of elephants appears consistent with the occurrence of different ancestral elephant species in the stratigraphic record.

little is known concerning the relative performance of the different methods. In this chapter, we review several recently developed methods designed to incorporate stratigraphic data into phylogenetic analysis. Although in this chapter we discuss stratigraphic data, except in very rare cases (e.g., Krings et al. 1997) fossils can be used only with morphological data.

CHANGING PERCEPTIONS OF THE IMPORTANCE OF FOSSILS

For about a century after the publication of *On the Origin of Species* (Darwin 1859), paleontologists were the main practitioners of phylogenetics. Many, perhaps most, biologists believed that fossil evidence was necessary to construct accurate evolutionary trees (e.g., Simpson 1961). However, with the introduction of explicit phylogenetic methods and of new sources of data, such as amino acid or DNA sequences, attitudes changed about the relevance of fossils in a phylogenetic analysis. Because fossils contain only a small fraction of the characters present in extant species, they came to be thought of as less useful for resolving relationships (Jefferies 1979; Hennig 1981; Patterson 1981; Rosen et al. 1981; Forey 1986).

In the late 1980s several groups rigorously examined the effect fossils have on phylogenetic analysis (see review by Donoghue et al. 1989). For example, Doyle and Donoghue (1987) examined the importance of fossils for seed plant phylogeny. Their general conclusion was that although the fossil data did not overturn a phylogeny constructed from data on living taxa, the interpretation of character evolution did change. Gauthier et al. (1988) examined the importance of fossils by studying amniotes. The results of their analysis suggested that the inclusion of fossils was critical. The phylogenetic tree constructed in the absence of fossils placed birds and mammals as sister taxa, whereas the inclusion of fossils resulted in a tree with Reptilia and mammals as sister taxa. Importantly, the phylogenetic tree proposed by Gauthier et al. (1988) was more consistent with the order of occurrence of fossil groups in the rock record. Moreover, the tree proposed by Gauthier et al. (1988) has been confirmed by collection of additional data (e.g., the majority of the genes in Hedges 1994). The critical fossils in the Gauthier et al. (1988) analysis turned out to be synapsid fossils. However, the critical fossils were neither the oldest nor the youngest included in the analysis.

Huelsenbeck (1991) performed simulation studies to tease apart the relationship between the temporal position of a fossil and its completeness. All else being equal, including fossils that were more ancestral was more useful to phylogenetic analysis than adding living taxa. Also, all else being equal, fossils that were more complete were better than fossils that preserved only a small fraction of characters. Interestingly, there were many situations in which the inclusion of an ancestral fossil that preserved only a small fraction of the characters was better than the addition of a living taxon to the analysis. These findings were consistent with the conclusions of Gauthier et al. (1988) and of Donoghue et al. (1989)—that the importance of fossils in a phylogenetic analysis is related to the combination of primitive and derived characters preserved in the fossils.

Fossils are important in phylogenetic analysis for what seems like the most obvious of reasons: fossils are remnants of the past, often the very distant past. Because of this, fossils often preserve unique combinations of characters that help resolve relationships. The purpose of this chapter is to discuss how the temporal uniqueness of fossils can be used in another manner: The temporal position of fossils can be used to test the fit of an estimated fossil to the rock record. Moreover, the stratigraphic position of fossils contains limited information on relationship. Although many readers may work with groups that have a poor fossil record or may work with molecular characters, for many problems it is important to consider the completeness of the stratigraphic record. For example, in several recent studies, the divergence times of major groups were examined using molecular data (for eubacteria/eukaryotes [Doolittle et al. 1996], metazoa [Wray et al. 1996], birds [Cooper and Penny 1997], mammalian orders and major lineages of vertebrates [Kumar and Hedges 1998]). In all of these studies, a molecular clock was assumed and at least one speciation time on the tree was considered known (to calibrate the clock). These researchers concluded that, for some major groups, the origin greatly predates the first fossils observed for the group. However, in none of these studies was uncertainty in the calibration times considered. To assess the uncertainty of the calibration times, models of fossil preservation are needed.

USING STRATIGRAPHIC DATA IN A PHYLOGENETIC ANALYSIS

Indices Measuring the Fit of Stratigraphy to Phylogeny

Several methods currently exist that are intended either to quantify the agreement of a phylogenetic tree with the stratigraphic record or to estimate phylogeny on the basis of information from the stratigraphic record. In this section, we focus on methods that measure the agreement between a phylogenetic tree and the stratigraphic record.

Spearman rank correlation. Gauthier et al. (1988) describe a method that investigates the degree of congruence between a phylogenetic tree and the stratigraphic record by comparing the "clade rank" with the "age rank" for each taxonomic group. The clade rank of a taxonomic group is defined as the number of nodes that must be traversed on a direct path from the root of the tree to the

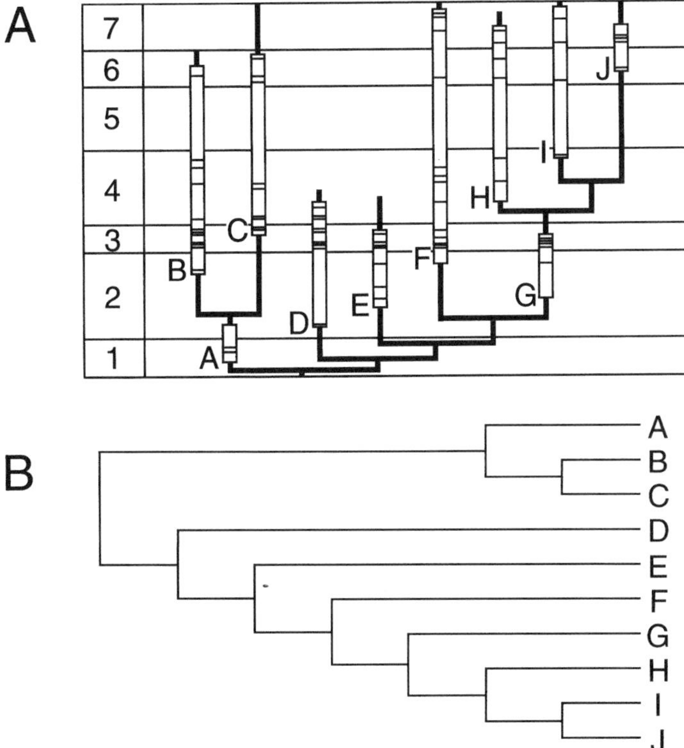

Figure 7.2. Sample data set. **(A)** The true evolutionary history of a group and its stratigraphic record. The true relationships and age ranges of the lineages are denoted by the thick solid lines. The observed age ranges are denoted by the rectangular boxes; within these boxes, the stratigraphic horizons at which a lineage is observed are denoted by dashes. **(B)** The cladogram that would be obtained if the relationships were correctly estimated.

branch-tip containing the group. The age rank of a taxonomic group is defined as the relative order of appearance of the group in the rock record (labeled from oldest to youngest). Gauthier et al. (1988) used Spearman's rank correlation (SRC; Sokal and Rohlf 1981) to evaluate the correlation of age and clade ranks for amniotes. The SRC coefficient, r_s, ranges in value from –1 to 1 and Gauthier et al. (1988) used it as a measure of the fit of the stratigraphic record to a phylogenetic tree. When $r_s = 1$, the order of appearance of groups agrees completely with their order of origination on the phylogeny, and when $r_s = -1$, these orderings are completely reversed. Figures 7.2 and 7.3 provide an example application of the method. Figure 7.2A depicts the actual evolutionary history of

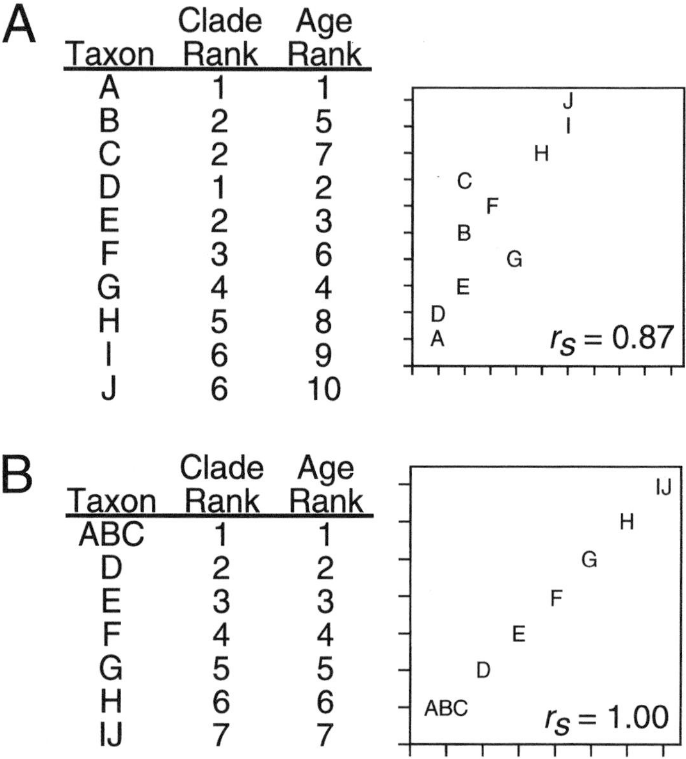

A

Taxon	Clade Rank	Age Rank
A	1	1
B	2	5
C	2	7
D	1	2
E	2	3
F	3	6
G	4	4
H	5	8
I	6	9
J	6	10

B

Taxon	Clade Rank	Age Rank
ABC	1	1
D	2	2
E	3	3
F	4	4
G	5	5
H	6	6
IJ	7	7

Figure 7.3. An example of Spearman rank correlation of clade and age ranks for the data from figure 7.2. **(A)** The relationship between clade and age rank according to the method of Gauthier et al. (1988). **(B)** The relationship between clade and age rank according to the method of Norell and Novacek (1992a).

a group of organisms, including the true age ranges and relationships of the species, as well as the observed age ranges. Figure 7.2B shows the cladogram that would result if the phylogeny of the 10 observed species were correctly estimated. Figure 7.3A shows the clade and age ranks of the 10 species, as well as the relationship between clade rank and age rank. For this example, $r_s = 0.87$.

The fact that the SRC coefficient is less than 1 for the tree in figure 7.2 indicates imperfect agreement between the phylogeny and the stratigraphic record. However, the imperfect fit results from the multiple clades of the same rank implied by the tree. Multiple clades of equal rank result when a tree is not maximally asymmetric (or pectinate) in shape. Figure 7.4 shows examples of maximally asymmetric and maximally symmetric trees. Norell and Novacek

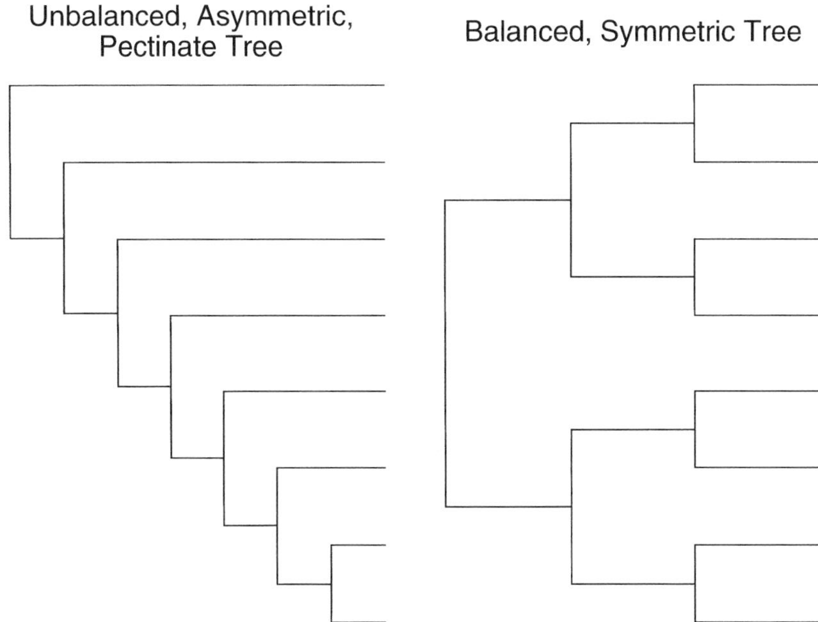

Figure 7.4. Examples of maximally balanced and unbalanced trees.

(1992a) modified the SRC approach to account for multiple clades of equal rank. Their approach is to pass through the tree, collapsing taxonomic groups with equal clade rank into a single taxonomic group and assigning that clade one rank. The oldest observed taxon from the collapsed group represents the age rank. Although this approach makes a correlation coefficient of 1 possible, it also discards potentially useful information because much of the stratigraphic information present in the collapsed clades is lost. Figure 7.3B illustrates clade and age ranks calculated by Norell and Novacek's (1992a) method. The SRC coefficient is designed to detect correlations between ordered pairs of observations. Although the ordered appearances of species in the fossil record may be viewed as observations, the clade ranks should not be treated as observations, as these are parameters inferred under an evolutionary model. As a consequence, the results for this index, at least as currently construed, do not have a rigorous statistical basis.

Stratigraphic consistency index. Huelsenbeck (1994) took a different approach to measuring the fit of the stratigraphic record to a phylogenetic tree. For every internal node of the phylogenetic tree (excluding the root node), the old-

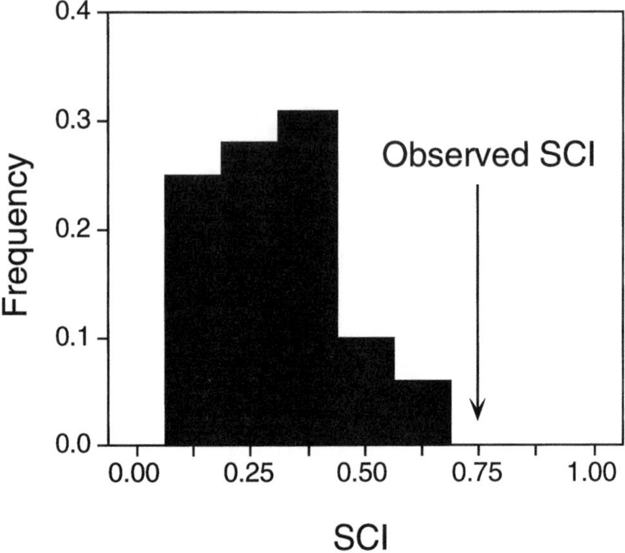

Figure 7.5. The distribution of stratigraphic consistency indices (SCI) expected under the null hypothesis of no association between phylogeny and the stratigraphic record for the data from figure 7.2.

est first occurrence in the fossil record of the species descended from the node is compared with the oldest first occurrence in the fossil record of its sister species. If the oldest first occurrence above the node is the same age as, or younger than, the oldest first occurrence of the sister taxon to the node, then the node is considered stratigraphically consistent. The stratigraphic consistency index (SCI) is intended to measure the proportion of stratigraphically consistent nodes.

The SCI varies between 0 and 1, and a low value may reflect disagreement between the fossil record and a phylogenetic tree or incomplete information on stratigraphic ranges. However, a high SCI value does not necessarily imply a good match between a phylogenetic tree and the stratigraphic record. For example, if all the fossil observations first occur at the same time instant, then SCI = 1. Huelsenbeck (1994) suggested that the observed SCI be compared to that expected under a null hypothesis of no association with the stratigraphic record for trees of the same shape; this can be done by permuting the ages among the tips of the tree many times. For each permutation, the SCI value is calculated. The proportion of the time that the observed SCI is exceeded in the permutations represents the probability of observing the SCI under the null hypothesis of random agreement between the stratigraphic record and phylogeny. Figure 7.5 shows the distribution of SCI values for the example from figure 7.2. The observed SCI value of 0.75 would not be expected to occur under the null hypothesis of no association between the stratigraphic record and phylogeny ($P < 0.01$).

Measures of gaps. Several methods aim to evaluate the amount of missing time (or gaps) implied by a phylogeny. Benton (1987) proposed the simple completeness metric (SCM), which is the ratio of the number of time intervals that a specific taxon was observed to the minimum number of time intervals during which the taxon must have existed. For example, if a taxon was observed in time intervals 1, 2, 3, 6, 9, and 10, then SCM = 0.6 (i.e., the taxon must have existed during at least 10 time intervals but was observed in only 6). For the five fossil species shown in figure 7.6A, the SCM values are 0.53, 1.00, 0.75, 0.55, and 0.60 for species A, B, C, D, and E, respectively. Norell and Novacek (1992a, 1992b) proposed the minimum implied gaps (MIG) method, intended to evaluate the minimum amount of missing time on a tree. MIG is calculated as the sum of the missing time for each lineage and its sister taxon on the tree. Benton (1994) modified the MIG measure. His relative completeness index (RCI) is equal to one minus the ratio of the MIG to the simple range lengths (SRL), where the SRL is a sum of the observed duration of each fossil taxon included in the analysis. A problem with such gap measures is that they are purely descriptive. In the absence of a null model, it is not obvious what SCM = 0.67 implies about the processes of evolution and fossil preservation.

Estimating Phylogeny from the Stratigraphic Record

Stratocladistics. The methods described above aimed to measure the congruence between the stratigraphic record and a phylogenetic tree. Fisher's (1982, 1988, 1991, 1992, 1994) stratocladistic method is intended to provide an estimate of phylogeny based on stratigraphic and morphological data. The method is similar in spirit to the gap indices discussed above. However, the method uses the amount of missing time (i.e., non-preservations of a species in the fossil record) implied by a tree to choose among alternative phylogenies. Missing time is measured as the number of unobserved "stratum crossings"—or instances that a taxon should have been present in a time interval (as implied by the phylogeny) but was not observed—so that each phylogenetic tree has an integer number associated with it that represents the minimum number of unobserved stratum crossings implied by that tree. The amount of missing time implied by a tree is its "stratigraphic debt" (Fisher 1994) and the tree requiring

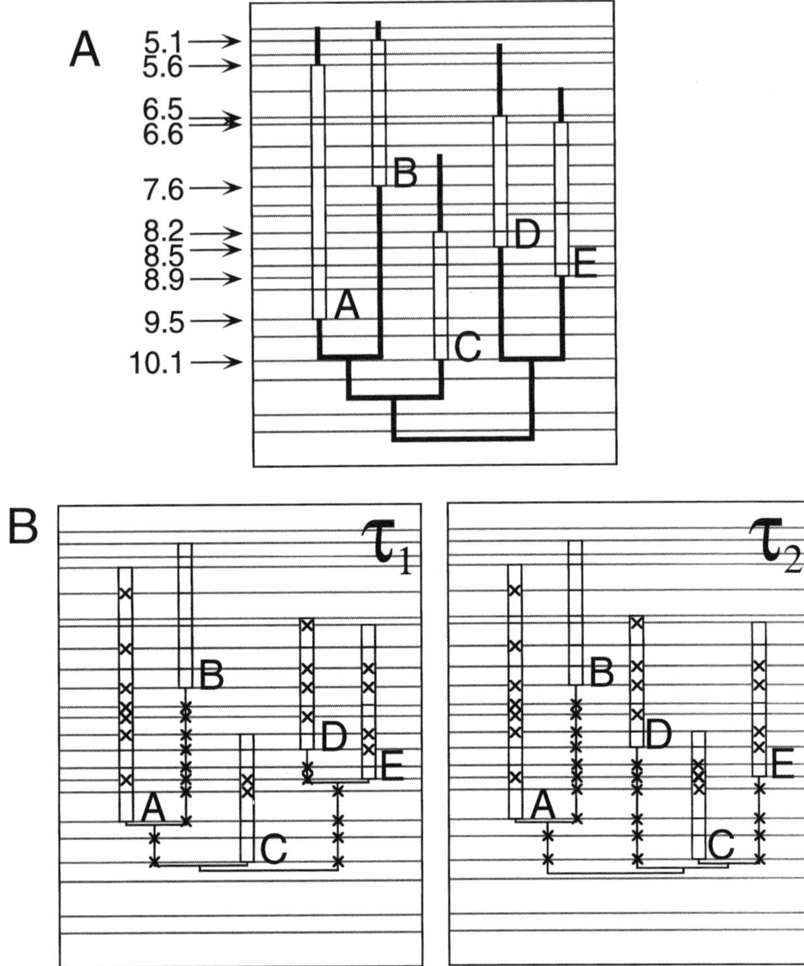

Figure 7.6. Consideration of stratigraphic debt. **(A)** The true genealogical relationships (thick solid lines) and observed stratigraphic ranges (rectangles and dashes therein) of five species. Numbers to the left represent millions of years. **(B)** Two alternative topologies, which have different levels of stratigraphic debt. Unobserved stratum crossings are denoted by "**x**."

the minimum amount of missing time is chosen as the best estimate of phylogeny in the absence of morphological information. An example calculation of stratigraphic debt is shown in figure 7.6B. Tree τ_1 implies a total of 33 unobserved stratum crossings, whereas tree τ_2 implies a total of 37 unobserved stratum crossings. On the basis of the stratigraphic record only, τ_1 is favored over

τ_2. If instead of measuring debt in terms of unobserved stratum crossings one uses absolute time, τ_1 implies 4.1 million years of missing time, whereas τ_2 implies 5.3 million years of missing time.

Gap methods using implied first and last fossil occurrences. Wagner (1995) proposed a method for choosing among phylogenetic trees that uses information from the first and last observed occurrence of a fossil species, as well as the total number of occurrences for each species. Wagner's method uses the 95% confidence limits on the first and last occurrences to find the tree that implies the fewest gaps (using the unobserved confidence intervals included in the range of a species). This tree is chosen as the best estimate of phylogeny. For a single taxon, the confidence intervals for the true first and last occurrence may be calculated by assuming that fossil preservation can be described as a Poisson process with rate λ (Strauss and Sadler 1989).

Maximum Likelihood Estimation of Phylogeny with Stratigraphic Data

Maximum likelihood is a standard statistical method for estimating the parameters of a model by using observed data generated under the model. An example of a maximum likelihood estimate is the familiar sample mean

$$\bar{x} = \frac{1}{n}\sum_{i=1}^{n} x_i,$$

which provides an estimate of the true mean, the parameter μ, on the basis of n independent observations, $\mathbf{x} = \{x_1, \ldots, x_n\}$, from a normal distribution, $N(\mu, \sigma^2)$. In an earlier study, we showed that a maximum likelihood approach can be used to estimate phylogenetic trees from stratigraphic data (Huelsenbeck and Rannala 1997). Fossil preservation is assumed to follow a Poisson process with preservation rate λ, as in Strauss and Sadler's (1989) method for establishing confidence intervals on the stratigraphic range of a single species. The maximum likelihood method uses information on the first and last occurrences of fossil taxa, as well as the total number of fossil occurrences for each taxon. The maximum likelihood approach is not restricted to a Poisson preservation model, however, and can be used with any stochastic model of fossil preservation. Likelihood ratio tests can be used to evaluate the relative fit of different models of fossil preservation to the observed data.

Models of fossil preservation. Several models of fossil preservation have been proposed. Here we review some of these models and discuss how they

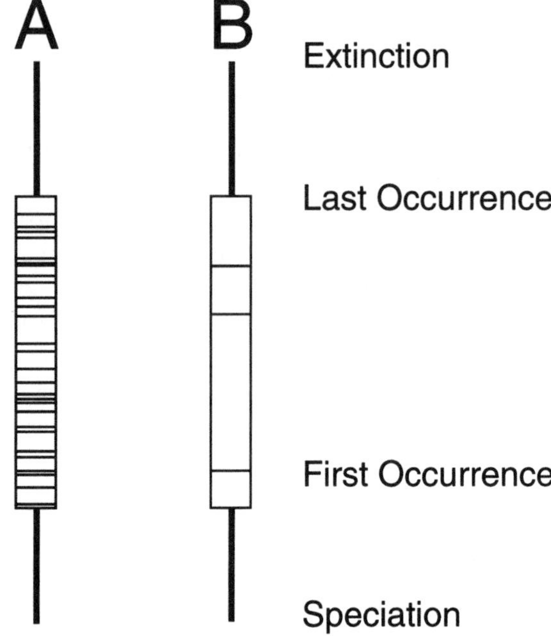

Figure 7.7. An example of two species with the same true range (the time between speciation and extinction) and the same observed range (the time between the first and last occurrences). However, species A was observed 30 times (including the first and last occurrence), whereas species B was observed only 5 times. Under a Poisson process of preservation, in which the preservation probability in any small interval of time, *dt,* is λdt, the probability of observing the data for species B is higher than that for species A if the preservation rate is the same for both species.

might be applied in a phylogenetic analysis using maximum likelihood methods. A Poisson process model of preservation assumes that in any instant of time, *dt,* the probability of a fossilization event occurring is λdt. The intuition underlying this model of fossil preservation is shown in figure 7.7. Here, two species have exactly the same speciation and extinction times. The observed first and last occurrence of each species is also the same. Species A has six times the number of observed preservation events as does species B, however. Hence, the probability of the actual taxon range is higher for species B than for species A; we would not expect such large gaps with no preservations for species A (as is the case at the ends of its range) because the larger number of observed preservations implies a high preservation rate for A.

The model of preservation specifies the probability of observing the fossil occurrences of species, given the actual speciation and extinction times and the preservation rate. This is denoted as $f(o_f, o_l, n \mid t_f, t_l, \lambda)$, where o_f is the observed first occurrence, o_l is the observed last occurrence, n is the number of occurrences, t_f is the time at which the species first arose, t_l is the time at which the species went extinct, and λ is the preservation rate. Conditional on t_f and t_l, the distribution of fossil preservations is uniform on the interval $[t_f, t_l]$. The joint probability distribution of the first and last fossil observations is the joint distribution of the smallest and largest order statistics of a uniform distribution:

$$f(o_f, o_l \mid n, t_f, t_l) = \left(\frac{1}{(t_l - t_f)}\right)^n n(n-1)(o_l - o_f)^{n-2}.$$

The probability distribution $f(n \mid t_f, t_l, \lambda)$ under the model is Poisson with parameter $\lambda(t_l - t_f)$ and the joint distribution $f(o_f, o_l, n \mid t_f, t_l, \lambda)$ is then

$$f(o_f, o_l, n \mid t_f, t_l, \lambda) = f(o_f, o_l \mid n, t_f, t_l) f(n \mid t_f, t_l, \lambda) = \frac{(o_l - o_f)^{n-2} \lambda^n e^{-\lambda(t_l - t_f)}}{(n-2)!}.$$

This equation gives the probability of a particular set of first and last occurrences, o_f and o_l, and a total of n occurrences, given the actual speciation and extinction times, t_f and t_l, and the preservation rate, λ. This equation assumes that $n \geq 2$. Huelsenbeck and Rannala (1997) provide formulae for the other two possible cases ($n = 1$ and $n = 0$).

The Poisson process model of fossil preservation, as formulated above, assumes that fossils can be dated with a high degree of accuracy and that the fossil preservation rate is constant through time and does not vary among lineages. Although the constraints of a constant preservation rate through time and of no rate variation among lineages can be relaxed, most paleontological data are not collected with accurate enough time resolution to implement the Poisson process model as described. Often, the paleontologist will have only partial information on the time intervals between fossil observations. We consider three cases in which the relative amount of information about the temporal distribution of fossils differs: (1) the duration of the time interval between observed first and last occurrences, o_f and o_l, and the number of observations in between the first and last occurrences are known, (2) the duration of each time interval (such as the geological period or epoch) and the presence or absence of a fossil lineage in each time interval are known ($n = 0$ or $n \geq 1$), and (3) the durations of the time intervals are roughly equal in length and the presence or absence of a fossil lineage in each time interval is known. The duration of the ith time interval is denoted d_i. In the first case, the information on the preservation of a lineage may be represented as a vector, $\mathbf{n} = \{n_i\}$, where n_i represents the number of observations of the lineage in the ith time interval (there are I time intervals in total). As an example, if a lineage was thought to exist during 11 time intervals ($I = 11$), the data might be coded $\mathbf{n} = \{0, 0, 2, 0, 0, 5, 0, 10, 112, 9, 0\}$. In this case, there are no observations of the lineage during intervals 1 and 2, two observations during interval 3, and so on. For the second and third cases, only the presence or absence of a species in each time interval is known. Hence, the stratigraphic data is coded as $\mathbf{n} = \{n_i\}$, where n_i is now a binary (0, 1) random variable. An example of preservation data coded in this form is $\mathbf{n} = \{0, 0, 1, 0, 0, 1, 0, 1, 1, 1, 0\}$, where 1 denotes the presence of the lineage in the ith time interval and 0 denotes its absence.

If the duration of the ith interval is known (d_i), then the probability under a Poisson model of preservation of observing n_i fossil horizons is

$$f(n_i \mid \lambda, d_i) = \frac{e^{-\lambda d_i}(\lambda d_i)^{n_i}}{n_i!},$$

where λ is the preservation rate. If only the presence or absence of a fossil lineage in a time interval is known (i.e., $n_i = 0$ or $n_i = 1$), then the probability of n_i under a Poisson model of preservation is

$$f(n_i \mid \lambda, d_i) = n_i + (-1)^{n_i} e^{-\lambda d_i}.$$

The probability of observing the distribution of fossil preservations among the I intervals is the product of the probabilities of observing the data in each interval, because the intervals are independent under a Poisson process:

$$f(\mathbf{n} \mid \lambda, \mathbf{d}) = \prod_{i=1}^{I} f(n_i \mid \lambda, d_i).$$

Foote and Raup (1996) considered a model of fossil preservation in which the probability of preservation during each time interval is p, regardless of the duration of the time interval. The probability of observing a given distribution of fossil presences ($n_i = 1$) or absences ($n_i = 0$) during I time intervals is then

$$\prod_{i=1}^{I} p^{n_i}(1-p)^{1-n_i}.$$

Foote and Raup (1996) derive the probability of observing a taxon range of t time intervals given that the true range was I intervals:

$$f(t|I) = \begin{cases} (1-p)^I, & t=0 \\ T(1-p)^{(I-1)}p, & t=1. \\ (I-t+1)(1-p)^{(I-t)}p^2, & t>1 \end{cases}$$

Maximum likelihood estimation. Assuming independent preservation among lineages, the likelihood of a specific phylogenetic tree is the product of the probabilities of observing the data for each of the s branches of a tree. If preservations among lineages are independent and follow a Poisson process, the likelihood is

$$L(\tau, \tilde{t}_f, \tilde{t}_l, \lambda | \tilde{o}_f, \tilde{o}_l, \tilde{n}) = \prod_{i=1}^{s} f(o_f^i, o_l^i, n^i | t_f^i, t_l^i, \lambda),$$

where the product is over the s lineages of a phylogenetic tree (τ), and

$$\begin{aligned} \tilde{o}_f &= \tilde{o}_f^{\lambda}, \ldots, o_f^s, \\ \tilde{o}_l &= o_l^{\lambda}, \ldots, o_l^s, \\ \tilde{t}_f &= \tilde{t}_f^{\lambda}, \ldots, \tilde{t}_f^s, \\ \tilde{t}_l &= \tilde{t}_l^{\lambda}, \ldots, \tilde{t}_l^s, \text{ and} \\ \tilde{n}_o &= n_o^{\lambda}, \ldots, n_o^s. \end{aligned}$$

The parameters of the model (branch lengths, topology, and preservation rate) are estimated by maximizing the likelihood function. The maximum likelihood tree maximizes the probability of observing the fossil data given the model of fossil preservation. The likelihood for the other models of preservation described above can be calculated similarly (i.e., by taking the product of the probabilities of observing the data over the s lineages).

As an example calculation of the likelihood of possible phylogenetic trees for stratigraphic data, consider the two trees of figure 7.6B. Table 7.1 shows the

Table 7.1

The observations (o_f, o_l, and n) and maximum likelihood estimates of speciation and extinction times (t_f and t_l, respectively) for tree τ_1 of figure 7.6B

Lineage	o_f	o_l	n	t_f	t_l
A	9.5	5.6	8	9.5	5.6
B	7.6	5.1	9	9.5	5.1
C	10.1	8.2	6	10.1	8.2
D	8.5	6.5	5	8.9	6.5
E	8.9	6.6	6	8.9	6.6
G_1	—	—	0	10.1	9.5
G_2	—	—	0	10.1	10.1
G_3	—	—	0	10.1	8.9

Tree τ_1 implies three ghost taxa, denoted G_1, G_2, and G_3.

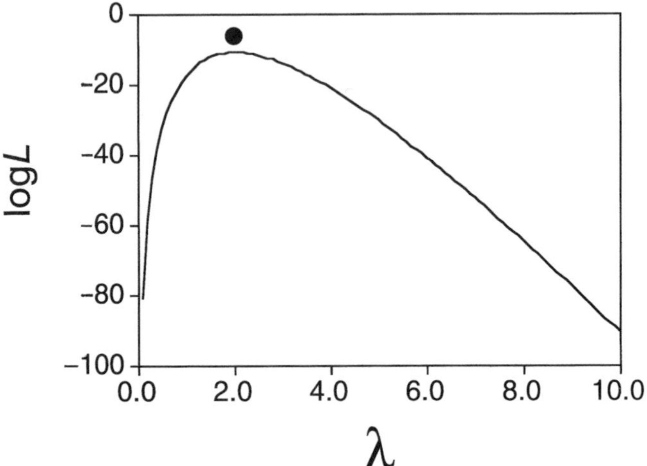

Figure 7.8. The log likelihood as a function of preservation rate (λ) for tree τ_1 from figure 7.6B. The maximum likelihood estimate of the fossil preservation rate is 2.02 (denoted by the large dot).

maximum likelihood estimates of speciation and extinction times for τ_1, as well as the information on the times of first and last occurrence and the total number of occurrences for each species. Note that tree τ_1 implies that there are three unobserved lineages ("ghost taxa" in Norell's [1992] terminology). The log likelihood, log*L,* of τ_1 is –10.72, whereas for τ_2 it is –12.88. Like the stratocladistic method, the maximum likelihood criterion favors τ_1. Although the maximum likelihood method provides an estimate of the phylogenetic tree and speciation and extinction times on that tree, it also provides an estimate of the preservation rate, λ. The maximum likelihood estimate of λ changes slightly depending on which tree is assumed. The estimate of λ is 2.02 for τ_1 and 1.90 for τ_2 (λ is in units of expected number of preservation events per million years). Figure 7.8 shows how the log likelihood changes as a function of λ for τ_1.

COMPARISON OF STRATIGRAPHIC METHODS

Most methods considered here for measuring the fit of the stratigraphic record to a phylogenetic tree are simple indices. These can be useful, but it is often not clear how a specific index value should be interpreted. For example, the SCI, by itself, is not very useful; that is, if someone were to state that the SCI for a particular data set was 0.5, should this be interpreted as a good or poor fit of the stratigraphic record to a tree? The same problem exists for other measures of stratigraphic fit (e.g., SRC, MIG, RCI). A potential solution is to compare the observed fit (using SCI, SRC, MIG, RCI, or some other method) to that expected if there were no correspondence between phylogeny and stratigraphy. One way to do this is to permute the observed stratigraphic ages among the tips of a phylogenetic tree (Huelsenbeck 1994). For each permutation, the index is recalculated and the proportion of the time that the observed value is exceeded in the permutations estimates the probability that the observed value would be generated given a random association of phylogeny with stratigraphy. Unfortunately, this approach has rarely been used to evaluate data.

Another problem with the various indices we have described is that their values may be strongly influenced by the shape of a tree (Siddall 1996). For example, for the SCI, maximally asymmetric trees can be made to more easily fit the stratigraphic record than can symmetrical trees. Caution must therefore be exercised when comparing indices among trees that have different shapes.

The remaining approaches that we have discussed—stratocladistics, Wagner's method, and the maximum likelihood approach—are all intended to estimate phylogenetic trees. Simulation analysis could be used to evaluate which of the methods performs best in estimating phylogeny. Such studies have not yet been performed, although we have investigated the statistical properties of the maximum likelihood method by means of simulations (Huelsenbeck and Rannala 1997). The maximum likelihood and stratocladistic methods should have similar performance because there is a strong relationship between the two methods; under a Poisson process of fossil preservation, the likelihood for a given phylogenetic tree is maximized when the amount of missing time implied by the tree is minimized (Huelsenbeck and Rannala 1997), and stratocladistics also seeks to minimize missing time. Apart from the accurate estimation of phylogeny, other criteria are also important in evaluating methods. One of the most important criteria is the ability to extend a method to address additional questions that may depend on phylogeny. Almost all of the methods discussed here can be modified to address additional questions concerning such processes as fossil preservation, support for different tree rootings, and the degree of phylogenetic agreement among diverse forms of data. However, maximum likelihood is unique among the methods so far developed in providing a unified set of procedures for testing hypotheses through the use of likelihood ratio tests.

COMBINING DATA

How can one best combine different sources of information, such as morphological data, stratigraphic data, or DNA sequences, to obtain an estimate of phylogeny? Fisher (1992) argued that the best overall estimate of phylogeny is the tree with the minimum overall length, as determined by adding the minimum number of character transformations implied by a tree (under a maximum parsimony criterion) to the minimum number of stratum crossings implied by that same tree overlaid on the fossil record. Fisher's method then allows various sources of data (morphological, molecular, and stratigraphic) to be combined in a phylogenetic analysis. MacClade 3.0 (Maddison and Maddison 1992) can be used to calculate stratigraphic debt for user-defined trees. Gingerich (1979) suggested another possible means of combining morphological and stratigraphic data—stratophenetics. The approach relies on the fine-scale resolution of the stratigraphic record and morphological comparisons among fossil taxa. Working from the base of the stratigraphic record upward, one views phenetically similar samples in different strata as being ancestors and descendants. Finally, the maximum likelihood method (Huelsenbeck and Rannala 1997) provides an alternative means of combining stratigraphic, morphological, and molecular data. If these data are independent, then their likelihoods, for a particular phylogeny, can be multiplied. As mentioned previously, the calculation of the likelihood for stratigraphic data under some simple models of fossil preservation is computationally simple. Similarly, calculating the likelihood of a tree for molecular data, such as DNA sequences, is easily done with existing methods (Felsenstein 1981a). Calculating the likelihood of morphological data is more difficult because few reasonable models exist for morphological evolution. A likelihood analysis can be performed for continuously varying characters by assuming a Brownian motion model of evolution (e.g., using CONTML; Felsenstein 1981b, 1995). This model has been applied to discrete morphological data with the traits treated as frequencies (Wiens 1998) and appears to behave similarly to frequency parsimony. Moreover, simulations suggest that CONTML performs well under a variety of conditions (Wiens and Servedio 1998).

The advantage of maximum likelihood methods over stratigraphic parsimony is that the weights given to the various sources of data are (1) proportional to their probabilities under the evolutionary models and (2) automatic, requiring no hidden assumptions on the part of the user (apart from the explicit assumptions specifying the model under which the analysis is performed). By contrast, the weight assigned to stratigraphic data by the stratocladistic method can be arbitrarily increased or decreased by considering more or less finely resolved stratigraphic units and has no obvious relationship to the probability of observing the data.

LIKELIHOOD RATIO TESTS WITH STRATIGRAPHIC DATA

There are a number of questions in paleontology that likelihood ratio tests can address. The ratio of the likelihoods calculated under two hypotheses,

$$\Lambda = \frac{\max[L(\text{Null Model} \mid \text{Data})]}{\max[L(\text{Alternative Model} \mid \text{Data})]},$$

is a measure of the tenability of one hypothesis over the other. Typically, likelihood ratio tests are constructed in such a way that the hypotheses considered are nested (i.e., the null hypothesis is a special case of the alternative hypothesis). In the case of nested hypotheses, $-2\log L$ is χ^2 distributed with q degrees of freedom, where q is the number of additional parameters under the alternative hypothesis. For many problems in phylogenetics, even though the hypotheses considered may appear to be nested, the χ^2 distribution does not closely approximate the null distribution. Monte Carlo simulation (or parametric bootstrapping) must then be used to approximate the null distribution of $-2\log L$ (Cox 1961, 1962; Felsenstein 1988; Goldman 1993). The parametric bootstrap procedure uses maximum likelihood estimates of model parameters, obtained under the null hypothesis, to simulate many data sets of the same size as the original, under the assumption that the null hypothesis (and estimated parameters) are true. For each simulated data set, the test statistic $-2\log L$ is calculated. The proportion of the time that the original value of $-2\log L$ is exceeded in the simulated data estimates the significance level of the test.

Congruence of Trees Estimated from Diverse Data

One possible application of likelihood ratio tests is to evaluate the congruence of phylogenetic trees estimated from diverse forms of data (Huelsenbeck and Bull 1996). For example, two possible types of data considered in paleontology are morphological characters for particular taxa and their stratigraphic occurrences. Alternatively, one might compare molecular data, such as DNA sequences, with stratigraphic occurrences. There are several reasons why one would want to test the congruence of different data partitions (see Huelsenbeck and Bull 1996). One reason is that one may want to evaluate the congruence of the stratigraphic data with an estimate of phylogeny based on morphological or molecular data. The null hypothesis assumes that the same phylogeny underlies both data sets. The alternative hypothesis relaxes the constraint that the same phylogeny underlies both data partitions. (This does not literally mean that different trees underly each data partition; rather, it is used to ask whether the different data sets are making significantly different statements about phylogeny.) The significance of the likelihood ratio test statistic, $-2\log L$, is determined through parametric bootstrapping (Huelsenbeck and Bull 1996). Rejection of the null hypothesis suggests that the data partitions support different phylogenetic trees and that this support is greater than would be expected from sampling error alone.

Often it will be difficult to perform the test of congruence as described above. For example, the likelihood of the morphological data may often be difficult to calculate due to the lack of an appropriate model describing the evolution of morphological characters. An alternative test might consider the fit of stratigraphic data to a tree constructed on the basis of morphology. This test assumes that the tree based on morphology is known without error. The null hypothesis assumes that this tree is correct and calculates the likelihood of the stratigraphic data given this tree. The likelihood of the stratigraphic data under the alternative hypothesis is calculated by allowing topology to be a free parameter; that is, the likelihood under the alternative hypothesis is the tree that maximizes the likelihood for the stratigraphic data. The null distribution of the likelihood ratio test statistic can be determined by parametric bootstrapping.

Other tests examining the concordance of a phylogenetic tree with the stratigraphic record that do not use likelihood ratios have been suggested (using Spearman's rank correlation—Gauthier et al. 1988, Norell and Novacek 1992a; using SCI—Huelsenbeck 1994; using SRC—Benton 1994). Very little can be said about the performance of these tests because no studies (e.g., simulations) have been done to evaluate the relative performance of these tests. However, it is known that likelihood ratio tests often perform well even when the hypotheses considered are composite (Rice 1995).

Testing Models of Fossil Preservation

Models of fossil preservation are important for several reasons. First, these models provide a basis for constructing confidence intervals of fossil species ranges over time. In fact, a statistical test for a mass extinction event has been proposed which relies on a Poisson process model of fossil preservations (Solow 1996). A model of fossil preservation also underlies the recent method of Foote (1997) for estimating extinction and preservation rates. Second, models of fossil preservation provide important information about the rate at which fossils are preserved. The field of taphonomy is concerned with the process of fossil preservation and a model-based approach might be used in this area to ask more detailed questions concerning the nature of the process of fossil preservation.

In as much as models of fossil preservation are important, it is also important to be able to improve preservation models. To test the fit of stratigraphic data to one model of fossil preservation versus another, one can use a likelihood ratio test. More complex models of fossil preservation than the Poisson process can be studied by introducing additional parameters into a model. For example, one could designate the Poisson process model typically used by paleontologists as the null model (Strauss and Sadler 1989). This model assumes that preservation rates are constant through time and identical across lineages. An alternative hypothesis might relax the constraint that preservation rate is constant through time or that rates are identical across lineages. Additional preservation parameters could then be estimated for different time intervals or for different lineages. The likelihood ratio test statistic calculated for the null and alternative models, $-2\log L$, is then compared to a χ^2 distribution with the appropriate degrees of freedom. The importance of the likelihood ratio test for models of fossil preservation is that it provides a framework for the gradual improvement of our current models.

ROOTING PHYLOGENETIC TREES

A powerful use of stratigraphic data is for rooting phylogenetic trees. Figure 7.9 shows an example of an unrooted four-taxon tree and the five possible rootings of the tree. The tree can be rooted along any one of the five branches. Phylogenies are most often rooted according to the outgroup criterion; species that

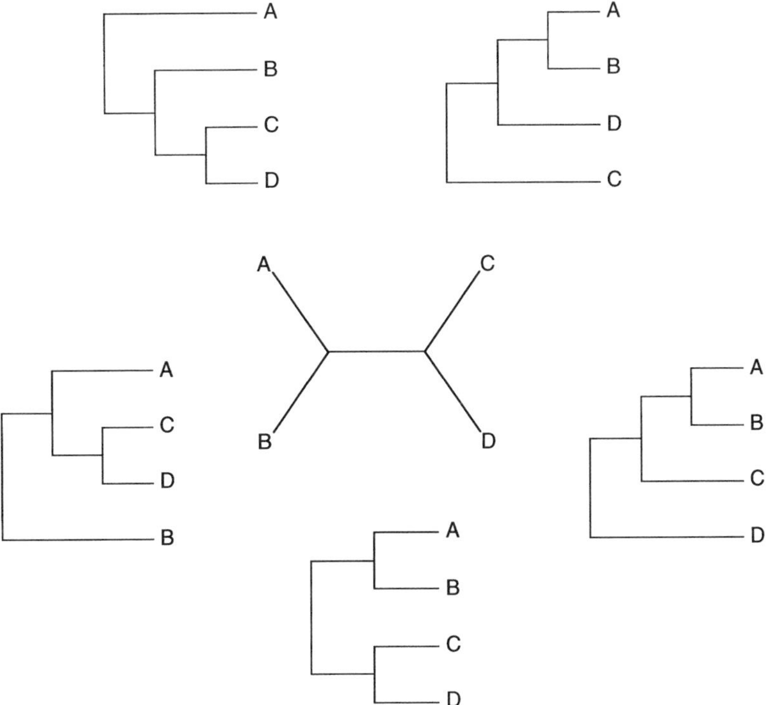

Figure 7.9. The unrooted and rooted trees for four species. The rooted trees are derived from the unrooted tree by considering each branch of the unrooted tree as a potential root.

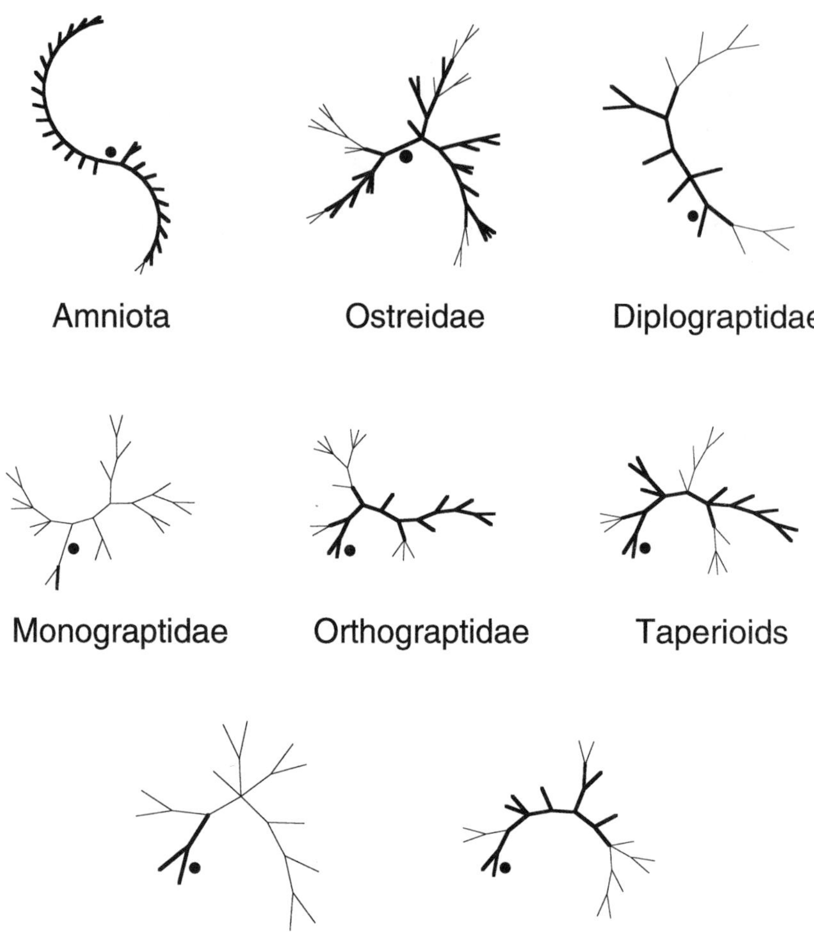

Figure 7.10. The branches chosen by the outgroup criterion (thin lines, rooted at the dot) and by the stratigraphic criterion (thick lines) on eight trees. Morphological data and analyses based on the outgroup criterion are from the following sources: Amniota, Gauthier et al. (1988); Ostreidae, Huelsenbeck (1992); Diplograptidae, Monograptidae, and Orthograptidae, Mitchell (1987); taperioids, Schoch (1989); Eleutherozoa, Smith (1988); peccaries, Wright (1993). Reanalyses of these data sets according to the stratigraphic criterion are from Huelsenbeck (1994).

are considered to be outside of the group of interest are included in the analysis to provide a root. Methods that do not rely on the outgroup criterion can also be used for rooting trees. In particular, rooted trees can be obtained from molecular data when these data fit a molecular-clock model. Methods have been proposed for polarizing characters on the basis of ontogeny or other criteria (see de Queiroz [1985] or Bryant [1991] for reviews of character polarity criteria). The stratigraphic record can also be used to root a phylogenetic tree (Huelsenbeck 1994). The stratigraphic rooting criterion considers each possible rooting position for an unrooted phylogenetic tree. The root position that provides the best fit to the stratigraphic record (under some optimality criterion) can be chosen as the root of the tree.

How well does the stratigraphic record root phylogenetic trees? Figure 7.10 shows eight trees that Huelsenbeck (1994) evaluated by using Fisher's (1992) stratocladistic method to estimate the optimal root(s). The root(s) indicated by the stratigraphic data were very close (or identical) to the root chosen by applying the outgroup criterion. Although the probability that the outgroup rooting position falls along the same region of the tree is often very high because of the large number of branches indicated as possible rooting positions by the stratigraphic criterion, when all of the studies are taken together, the concordance is remarkable. The congruence of the rooting position chosen by two independent methods strengthens our confidence of the accuracy of both methods for rooting trees. The tree root may also be chosen by using the maximum likelihood criterion (see Huelsenbeck and Rannala 1997).

CONCLUSIONS

Although stratigraphic data have been incorporated into phylogenetic analyses since at least the 1860s, such information is only infrequently considered by modern systematists. Quite apart from phylogenetic information arising from unique combinations of character states (Gauthier et al. 1988), fossils provide an additional source of information in phylogenetics. Minimally, stratigraphic data can be used to (1) estimate phylogeny, (2) examine the congruence between the stratigraphic record and phylogeny, (3) examine models of fossil preservation, and (4) root phylogenetic trees. In this chapter, we have attempted to outline some of the approaches that have been suggested for incorporating stratigraphic data into a phylogenetic analysis. We show that all of the fossil applications outlined above can be performed by using a unified set of tools arising from the well-known statistical procedure of likelihood analysis. It is our hope that this discussion will motivate systematists to further explore the potential uses of stratigraphic data and likelihood methods in phylogenetic analysis.

ACKNOWLEDGMENTS

This research was supported by a Miller postdoctoral fellowship awarded to J.H. and a Natural Sciences and Engineering Research Council (NSERC) of Canada postdoctoral fellowship awarded to B.R.

REFERENCES

Benton, M. J. 1987. Mass extinctions among families of non-marine tetrapods: The data. Mémoires de la Société Géologique de France, Nouvelle Série 150: 21–32.

Benton, M. J. 1994. Paleontological data and identifying mass extinctions. Trends in Ecology and Evolution 9: 181–185.

Benton, M. J. 1995. Testing the time axis of phylogenies. Philosophical Transactions of the Royal Society of London B 349: 5–10.

Benton, M. J., and G. W. Storrs. 1996. Diversity in the past: Comparing cladistic phylogenies and stratigraphy. Pages 19–40 *in* Aspects of the Genesis and Maintenance of Biological Diversity (ed. M. E. Hochberg, J. Clobert, and R. Barbault). Oxford University Press, Oxford.

Brochu, C. A. 1997. Morphology, fossils, divergence timing, and the phylogenetic relationships of *Gavialis*. Systematic Biology 46: 479–522.

Bryant, H. 1991. The polarization of character transformations in phylogenetic systematics: Role of axiomatic and auxiliary assumptions. Systematic Zoology 40: 433–445.

Cooper, A., and D. Penny. 1997. Mass survival of birds across the Cretaceous–Tertiary boundary: Molecular evidence. Science 275: 1109–1113.

Cox, D. R. 1961. Tests of separate families of hypotheses. Proceedings of the Fourth Berkeley Symposium. University of California Press, Berkeley.

Cox, D. R. 1962. Further results on tests of separate families of hypotheses. Journal of the Royal Society of London B 24: 406–424.

Darwin, C. 1859. On the Origin of Species by Means of Natural Selection. J. Murray, London.

de Queiroz, K. 1985. The ontogenetic method for determining character polarity and its relevance to phylogenetic systematics. Systematic Zoology 34: 280–299.

Donoghue, M. J., J. A. Doyle, J. Gauthier, A. G. Kluge, and T. Rowe. 1989. The importance of fossils in phylogeny reconstruction. Annual Review of Ecology and Systematics 20: 431–460.

Doolittle, F. F., D.-F. Feng, S. Tsar, G. Cho, and E. Little. 1996. Determining divergence times of the major kingdoms of living organisms with a protein clock. Science 271: 470–477.

Doyle, J. A., and M. J. Donoghue. 1987. The importance of fossils in elucidating seed plant phylogeny and macroevolution. Review of Paleobotany and Palynology 50: 63–95.

Felsenstein, J. 1981a. Evolutionary trees from DNA sequences: A maximum likelihood approach. Journal of Molecular Evolution 17: 368–376.

Felsenstein, J. 1981b. Evolutionary trees from gene frequencies and quantitative characters: Finding maximum likelihood estimates. Evolution 35: 1229–1241.

Felsenstein, J. 1988. Phylogenies from molecular sequences: Inference and reliability. Annual Review of Genetics 22: 521–565.

Felsenstein, J. 1995. PHYLIP (Phylogeny Inference Package), version 3.57c. Distributed by the author. Department of Genetics, University of Washington, Seattle.

Fisher, D. C. 1982. Phylogenetic and macroevolutionary patterns within the Xiphosurida. North American Paleontological Convention III, Proceedings 1: 175–180.

Fisher, D. C. 1988. Stratocladistics: Integrating stratigraphic and morphologic data in phylogenetic inference. Geological Society of America, Abstracts with Program 20: A186.

Fisher, D. C. 1991. Phylogenetic analysis and its application in evolutionary paleobiology. Pages 103–122 *in* Analytical Paleobiology (ed. N. L. Gilinsky and P. W. Signor). Short Courses in Paleontology, No. 4. Paleontological Society.

Fisher, D. C. 1992. Stratigraphic parsimony. Pages 124–129 *in* MacClade: Analysis of Phylogeny and Character Evolution (ed. W. P. Maddison and D. R. Maddison). Sinauer Associates, Sunderland, Mass.

Fisher, D. C. 1994. Stratocladistics: Morphological and temporal patterns and their relation to phylogenetic process. Pages 133–171 *in* Interpreting the Hierarchy of Nature—from Systematic Patterns to Evolutionary Theories (ed. L. Grande and O. Rieppel). Academic Press, Orlando, Fla.

Foote, M. 1997. Estimating taxonomic durations and preservation probability. Paleobiology 23: 278–300.

Foote, M., and D. M. Raup. 1996. Fossil preservation and the stratigraphic ranges of taxa. Paleobiology 22: 121–140.

Forey, P. L. 1986. Relationships of lungfishes. Journal of Morphology, Supplement 1: 75–91.

Gaudry, A. 1862. Animaux fossiles et géologie de l'attique d'aprés les recherches faites en 1855–56 et 1860 sous les auspices de l'Académie des Sciences, Paris. F. Savy, Paris.

Gauthier, J., A. G. Kluge, and T. Rowe. 1988. Amniote phylogeny and the importance of fossils. Cladistics 4: 105–209.

Gingerich, P. D. 1979. The stratophenetic approach to phylogeny reconstruction in vertebrate paleontology. Pages 41–77 *in* Phylogenetic Analysis and Paleontology (ed. J. Cracraft and N. Eldredge). Columbia University Press, New York.

Goldman, N. 1993. Statistical tests of models of DNA substitution. Journal of Molecular Evolution 36: 182–198.

Harper, C. 1976. Phylogenetic inference in paleontology. Journal of Paleontology 50: 180–193.

Hedges, S. B. 1994. Molecular evidence for the origin of birds. Proceedings of the National Academy of Sciences, U.S.A. 91: 2621–2624.

Hennig, W. 1981. Insect Phylogeny. Wiley, New York.

Huelsenbeck, J. P. 1991. When are fossils better than extant taxa in phylogenetic analysis? Systematic Zoology 40: 458–469.

Huelsenbeck, J. P. 1992. Oyster phylogeny: Fossils and confidence. Master's thesis, University of Texas, Austin.

Huelsenbeck, J. P. 1994. Comparing the stratigraphic record to estimates of phylogeny. Paleobiology 20: 470–483.

Huelsenbeck, J. P., and J. J. Bull. 1996. A likelihood ratio test to detect conflicting phylogenetic signal. Systematic Biology 45: 92–98.

Huelsenbeck, J. P., and B. Rannala. 1997. Maximum likelihood estimation of phylogeny using stratigraphic data. Paleobiology 23: 174–180.

Jefferies, R. P. S. 1979. The origin of chordates: A methodological essay. Pages 443–477 *in* The Origin of Major Invertebrate Groups (ed. M. R. House). Academic Press, London.

Krings, M., A. Stone, R. W. Schmitz, H. Krainitzki, M. Stoneking, and S. Paabo. 1997. Neandertal DNA sequences and the origin of modern humans. Cell 90: 10–30.

Kumar, S., and S. B. Hedges. 1998. A molecular timescale for vertebrate evolution. Nature 392: 917–920.

Maddison, W. P., and D. R. Maddison. 1992. MacClade: Analysis of Phylogeny and Character Evolution, version 3.0. Sinauer Associates, Sunderland, Mass.

Marshall, C. R. 1990. Confidence intervals on stratigraphic ranges. Paleobiology 16: 1–10.

Mitchell, C. E. 1987. Evolution and phylogenetic classification of the Diplograptacea. Paleontology 30: 353–405.

Norell, M. 1992. Taxic origin and temporal diversity: The effect of phylogeny. Pages 89–118 *in* Extinction and Phylogeny (ed. M. J. Novacek and Q. D. Wheeler). Columbia University Press, New York.

Norell, M., and M. Novacek. 1992a. The fossil record and evolution: Comparing cladistic and paleontologic evidence for vertebrate history. Science 255: 1690–1693.

Norell, M., and M. Novacek. 1992b. Congruence between superpositional and phylogenetic patterns with fossil records. Cladistics 8: 319–337.

Patterson, C. 1981. Significance of fossils in determining evolutionary relationships. Annual Review of Ecology and Systematics 12: 195–223.

Rice, J. A. 1995. Mathematical Statistics and Data Analysis. Duxbury Press, Belmont, Calif.

Rosen, D. E., P. L. Forey, B. G. Gardiner, and C. Patterson. 1981. Lungfishes, tetrapods, paleontology, and plesiomorphy. Bulletin of the American Museum of Natural History 167: 159–267.

Schoch, R. M. 1989. A review of the taperioids. Pages 298–320 *in* The Evolution of Perissodactyls (ed. D. R. Prothero and R. M. Schoch). Oxford University Press, New York.

Siddall, M. 1996. Stratigraphic consistency and the shape of things. Systematic Biology 45: 111–115.

Simpson, G. G. 1961. Principles of Animal Taxonomy. Columbia University Press, New York.

Smith, A. B. 1988. Patterns of diversification and extinction in Early Paleozoic echinoderms. Paleontology 31: 799–828.

Sokal, R. R., and F. J. Rohlf. 1981. Biometry, 2nd edition. W. H. Freeman, New York.

Solow, A. R. 1996. Tests and confidence intervals for a common upper endpoint in fossil taxa. Paleobiology 22: 406–410.

Strauss, D., and P. M. Sadler. 1989. Classical confidence intervals and Bayesian probability estimates for ends of local taxon ranges. Mathematical Geology 21: 411–427.

Wagner, P. J. 1995. Stratigraphic tests of cladistic hypotheses. Paleobiology 21: 153–178.

Wiens, J. J. 1998. Testing phylogenetic methods with tree-congruence: Phylogenetic analysis of polymorphic morphological characters in phrynosomatid lizards. Systematic Biology 47: 411–428.

Wiens, J. J., and M. R. Servedio. 1998. Phylogenetic analysis and intraspecific variation: Performance of parsimony, distance, and likelihood methods. Systematic Biology 47: 228–253.

Wray, G., J. S. Levinton, and L. Shapiro. 1996. Molecular evidence for deep Precambrian divergences among metazoan phyla. Science 274: 568–573.

Wright, D. 1993. Evolution of sexually dimorphic characters in peccaries (Mammalia, Tayassuidae). Paleobiology 19: 52–70.

KEVIN DE QUEIROZ

8

LOGICAL PROBLEMS ASSOCIATED WITH INCLUDING AND EXCLUDING CHARACTERS DURING TREE RECONSTRUCTION AND THEIR IMPLICATIONS FOR THE STUDY OF MORPHOLOGICAL CHARACTER EVOLUTION

One of the driving forces behind a renewed interest in phylogeny is the realization that a phylogenetic context is critical for the study of a wide variety of evolutionary questions (e.g., Eldredge and Cracraft 1980; Felsenstein 1985; O'Hara 1988; Donoghue 1989; Funk and Brooks 1990; Brooks and McLennan 1991; Harvey and Pagel 1991; Maddison and Maddison 1992; Eggleton and Vane-Wright 1994; Harvey et al. 1995). Of particular interest are questions about character evolution and adaptation—an area of investigation now commonly known as comparative biology. A prominent and unresolved problem in comparative biology is whether it is appropriate to include certain characters as part of the data used to reconstruct a phylogenetic tree when the evolution of those same characters is the subject of investigation. One widespread belief is that including the characters of interest is logically circular and that independent phylogenies (i.e., ones based on data other than the characters under investigation) are needed to properly analyze character evolution (e.g., Coddington 1988; Carpenter 1989; Lauder and Liem 1989; Olmstead 1989; Sytsma 1990; Brooks and McLennan 1991; McKey 1991; Vane-Wright et al. 1992; Hedges and Maxson 1996). An alternative and also widespread opinion is that the best inferences about character evolution must be based on the best estimates of phylogeny, which must in turn be based on all of the available evidence—that is, including the characters of interest (e.g., Kluge 1989; Donoghue and Sanderson 1992; Maddison and Maddison 1992; Swofford and Maddison 1992; Deleporte 1993; A. de Queiroz and Wimberger 1993; McKitrick 1993; Lee 1997). Because the characters of interest are often morphological, the issue of whether to include or exclude those characters has become entangled with another debate about the use of morphological versus molecular data in phylogeny reconstruction (for reviews of the latter, see Hillis 1987; Patterson 1987; chapter 1 of this book). Not surprisingly, authors who favor the use of molecular data for reconstructing phylogenetic trees commonly advocate excluding the characters of interest—particularly when those characters are morphological—and optimizing them on trees reconstructed entirely from molecular data (e.g., Hedges and Maxson 1996).

In this chapter, I analyze the practices of including and excluding the characters of interest during the reconstruction of phylogenetic trees in relation to the logic of analyzing character evolution in a phylogenetic context. I explain part of the controversy about whether to include or exclude characters as a manifestation of alternative perspectives on the general nature of inferences. I argue that even within the context of a perspective that emphasizes inferential caution, both including and excluding the characters of interest can lead to logical problems, which implies that although there is a valid concern about including the characters of interest during tree reconstruction, that concern cannot be eliminated by a simple rule to always exclude those characters. In light of these conclusions, I make both general and specific methodological recommendations about including and excluding characters that can be used to increase analytical rigor in studies of character evolution. Although based largely on an earlier article (K. de Queiroz 1996), this chapter extends my earlier analysis in several ways, including considerations of how the controversy about including versus excluding the characters of interest relates to the controversy about morphological versus molecular data, to prior assessments of homology versus analogy, and to different general classes of analyses. I have also expanded and modified the discussions of logical problems and methodological recommendations, including several examples from empirical studies.

DIFFERENT GENERAL PERSPECTIVES

The controversy about whether it is preferable to include or exclude the characters of interest stems, at least in part, from alternative general perspectives concerning the nature of inferences about character evolution. The different general perspectives can be characterized as attempting to make the inference that is most likely correct versus attempting to make an inference that is unlikely to be incorrect. These two perspectives are distinct—that is, they are not merely different ways of describing the same concern. Seeking the inference or estimate that is the most likely to be correct is an inherently bold or liberal approach in that magnitude of the difference between the best and second best estimate (as judged by some optimality criterion) is irrelevant. The best estimate is still the best estimate regardless of how much better it is than the next best estimate. A statistical example is point estimation, in which the inference or estimate that maximizes or minimizes some quantity is taken as the best estimate regardless of the magnitude of the difference between its value and those of alternative estimates. In this context, it makes sense to include all relevant characters in the data used to reconstruct the phylogenetic tree—even the characters of interest. To the extent that the information in the characters of interest is not predominantly misleading, those characters provide evidence about the phylogenetic relationships of the taxa in which they occur and thus also about their own evolution. Excluding that information can be expected to decrease the ability of any tree-based analytical method to yield the correct inference.

In contrast, seeking an inference or estimate that is unlikely to be incorrect is an inherently cautious or conservative approach in that the magnitude of the difference between one estimate and another is crucial. An estimate that is only marginally better than an alternative is much more likely to be incorrect than one that is substantially better, and a cautious researcher may prefer to abstain from making an inference when the best estimate has more than a very low probability of being incorrect. A statistical example is significance testing, which adopts the convention of rejecting the null hypothesis only when the data have a very low probability (<0.05) if the null hypothesis is true; but by that same convention, necessitates that in many cases one will be unable to reject the null hypothesis (i.e., make an inference) even though it is false. In this context, it sometimes makes sense to exclude the characters of interest from the data used to reconstruct the phylogenetic tree (see below). To the extent that including the characters of interest influences the outcome of an analysis in one direction or another, that practice can be expected—at least in some cases—to compromise the cautious or conservative nature of the inference.

Although these alternative perspectives lead researchers to opposite conclusions about whether to include or exclude the characters of interest, the perspectives themselves are complementary rather than contradictory. There is no contradiction in recognizing that the inference that is most likely correct may still have a relatively high probability of being incorrect—that the single best estimate of a parameter is not necessarily a conservative estimate. Moreover, there is no reason to question the validity of either general perspective; both reflect legitimate goals and concerns. In the rest of this chapter, however, I will (for the most part) restrict my discussion to the general perspective that emphasizes the avoidance of incorrect inferences, because this is the context within which concerns about including the characters of interest are most relevant.

LOGICAL FALLACIES AND RELATED INFERENTIAL PROBLEMS

Among authors who advocate excluding the characters of interest, there is a widespread opinion that including those characters is tantamount to fallacious reasoning (e.g., Felsenstein 1985; Coddington 1988; Olmstead 1989; Sytsma 1990; McKey 1991; Vane-Wright et al. 1992; Sillén-Tullberg and Møller 1993; Block and Finnerty 1994; Meyer et al. 1994; Hedges and Maxson 1996). However, just because a practice can compromise an inference does not mean that it necessarily will in all cases. I next examine the issue of logical fallacies in greater detail to show that previous statements about the problem have been oversimplified. I do so by applying some general ideas about inferential logic to the specific case of character evolution, arguing that the existence of logical problems in such studies is dependent on how the assumptions inherent in specific methods or practices bear on the specific inferences or conclusions of an analysis. This perspective is implicit in several earlier papers (e.g., Maddison 1990; Harvey and Purvis 1991; Armbruster 1992, 1993; Maddison and Maddison 1992; Swofford and Maddison 1992; Deleporte 1993; A. de Queiroz and Wimberger 1993) but has not previously been presented as a detailed and explicit argument.

Arguing in a Circle

The problem of including the characters of interest has most commonly been characterized as an example of logical circularity, or at least potential circularity (e.g., Felsenstein 1985; Coddington 1988; Olmstead 1989; Sytsma 1990; McKey 1991; Armbruster 1992, 1993; Donoghue and Sanderson 1992; Vane-Wright et al. 1992; A. de Queiroz and Wimberger 1993; Sillén-Tullberg and Møller 1993; Block and Finnerty 1994; Meyer et al. 1994; Hedges and Maxson 1996). Although there are precedents for this characterization, it is also potentially misleading. In the following discussion, I adopt the terminology of the *Oxford English Dictionary* (Second Edition, Oxford University Press, 1989), which uses different and descriptive terms to distinguish between two different, though related, classes of logical fallacies.

The fallacy termed "arguing or reasoning in a circle" (also "vicious circle" or "circular reasoning") is an invalid form of argument in which a proposition is used to establish a conclusion, and afterwards the proposition is justified by means of the very conclusion that it was previously used to establish. The description of the argument as circular derives from the fact that proposition and conclusion depend on one another in such a way that the inferential sequence has no clear starting point. A phylogenetic example would be using a tree reconstruction method that minimizes instances of homoplasy (e.g., parsimony), obtaining a result in which homoplasy is rare, and then using that result to justify use of the method that minimizes homoplasy.

The practice (proposition) in question here is that of including certain characters as part of the data used to estimate a phylogenetic tree, which then serves as the basis for inferences about the evolution of those same characters. In this case, a circular argument would be one in which a conclusion reached in an analysis that included the characters of interest was then used to justify the very practice of including those characters. This is not, however, the kind of logical problem about which comparative biologists have been concerned. In attempting to avoid circularity by excluding the characters of interest, comparative biologists have been concerned about their conclusions regarding character evolution—not about arguments or justifications for including the characters of interest. In other words, they have been concerned only with the validity of using a particular practice to arrive at a result or conclusion rather than with the full circle that then uses the result or conclusion to justify the original practice. Strictly speaking, this is not arguing in a circle (see also Maddison and Maddison 1992; Swofford and Maddison 1992).

Begging the Question

The fallacy termed "begging the question" (also "*petitio principii*") is more relevant to the issue of including the characters of interest in that it bears on the logic of establishing conclusions (i.e., as opposed to justifying initial propositions). To beg the question is to take for granted the matter in dispute or, more specifically, to assume the conclusion in the premises of the argument. A phylogenetic example would be reconstructing phylogenies using a method that explicitly or implicitly assumes evolutionary rate constancy among lineages (e.g., phenetic clustering) and then concluding, based on those phylogenies, that rates of evolution among lineages have been constant.

Strictly speaking, most analyses of character evolution that include the characters of interest during tree reconstruction do not constitute examples of begging the question. Including the characters of interest will seldom precisely entail whatever conclusion is ultimately reached concerning the evolution of those characters, though it is at least possible to contrive an example in which it does (e.g., arguing for the single origin of a particular character when that character was the only datum used to reconstruct, using parsimony, the phylogenetic tree upon which the conclusion of a single origin is based). On the other hand, less extreme influences on the outcome of at least some types of analyses regarding character evolution are expected to occur more commonly (see below). Therefore, in the context of an approach that favors inferential caution, there is at least a potential problem with including the characters of interest. That problem is conceptually related to the fallacy of begging the question, the difference being that the premises exert an undesirable influence on the conclusions without fully assuming them.

Inappropriately Influencing the Analysis

Various practices can systematically influence an analysis so that the analysis tends to favor a particular outcome or result. If the result that tends to be favored by a particular method or practice is the same one obtained in an analysis using that method, then a logical problem exists—namely, it is ambiguous whether the result is attributable to the data per se or whether it is an artifact of the methodological influence. I call this problem "inappropriately influencing the analysis." It does not require that the practice (premise) in question fully assumes the result (conclusion) but only that it tends to favor that result by influencing the analysis in that direction. Begging the question refers to a situation in which the practice in question makes a particular result inevitable and thus can be viewed as the most extreme form of an inappropriate influence.

The example used above to illustrate the fallacy of circular reasoning includes an example of an inappropriate influence (which nonetheless constitutes only part of the full circle illustrated in that argument). If one is going to conclude that homoplasy is rare based on the reconstruction of character state transformations on a phylogenetic tree, then the use of tree (and character state) reconstruction methods based on parsimony introduces an inappropriate influence. Although parsimony methods do not preclude the possibility of obtaining results in which homoplasy is inferred to be common (Farris 1983), they minimize and thus systematically underestimate homoplasy, which biases the analysis in favor of the conclusion that homoplasy is rare.

Problematic and Nonproblematic Influences

Not all systematic influences are logically problematic. Consider, once again, the use of parsimony methods in evaluating the commonness of homoplasy, except this time with the finding that homoplasy is common. The same bias exists as in the previous example, but this time it is not the source of inferential ambiguity. Because the method biases the analysis in favor of the inference that homoplasy is rare, obtaining the result that homoplasy is common cannot be attributed to the methodological influence. Indeed, given that the method is biased against the finding that homoplasy is common, the analysis would have to be considered conservative with respect to that result. The influence is unproblematic, and it might even be considered appropriate or desirable, given the results.

Generalizing from the previous examples, we can conclude that the appropriate versus inappropriate nature of a given influence depends on the context. An influence that tends to favor a particular hypothesis is problematic and thus inappropriate if that hypothesis is supported by the results of an analysis containing the influence in question. But an identical influence is nonproblematic and appropriate, in the sense of strengthening the inference, if the hypothesis favored by the influence is contradicted and thus an alternative hypothesis is supported by the results. The concepts of appropriate and inappropriate influences are closely related to the concepts of conservative and liberal tests. An appropriate influence makes a test more conservative, whereas an inappropriate one renders a test more liberal. In any case, the critical issue in classifying a systematic methodological influence as appropriate or inappropriate is whether the results are potentially attributable to that influence, because that is what compromises the reliability of the inference.

A systematic influence can also be considered nonproblematic if it is neutral with respect to the hypotheses being evaluated. If the influence in question does not favor any of the hypotheses under consideration, then it causes no inferential ambiguity regardless of which hypothesis is supported by the results. For example, the use of tree reconstruction methods based on parsimony would not be problematic with regard to the question of whether rates of evolution have been constant or variable among lineages. Insofar as those methods bias the results in terms of the frequency of homoplasy rather than the distribution of character transformations among lineages, they do not bias the results in favor of either constant or variable rates.[1]

In short, the classification of a given influence as problematic (inappropriate) or nonproblematic (appropriate or neutral) depends on the specific circum-

---

[1]On the other hand, use of specific parsimony-based optimization methods, such as accelerated versus delayed transformation (Swofford and Maddison 1987), might well bias inferences about evolutionary rates in individual lineages. It should also be noted that although certain patterns of rate inequality may cause parsimony methods to construct the incorrect topology (Felsenstein 1978), the problem does not result from rate inequality per se but rather from certain patterns of branch length inequality (Hendy and Penny 1989). Moreover, the patterns of branch length inequality in question do not necessarily prevent parsimony methods from identifying rate inequalities, though in cases of "long edge [branch] attraction" (Hendy and Penny 1989), the lengths of long branches will be underestimated and those of certain short branches will be overestimated.

stances of an analysis. In particular, it depends on how that influence affects the results of the analysis with respect to the specific hypothesis under consideration. Because of this context dependence, the same influence can be appropriate in one situation, neutral in another, and inappropriate in yet another. Consequently, it is simplistic to reject a given method or practice without considering the specific relationship between the systematic influences inherent in that method and the hypothesis that is supported by the results of an analysis in which it is employed, and it is almost certainly incorrect to think that a given method or practice will be logically problematic in all cases.

PROBLEMS ASSOCIATED WITH INCLUDING (AND EXCLUDING) THE CHARACTERS OF INTEREST

In the context of the analysis presented above, it should not be surprising that including the characters of interest—or, for that matter, excluding them—leads to logical problems in some cases but not in others. The existence of a logical problem depends on how each practice bears on the hypotheses under consideration. Consider, for example, a study that seeks to assess whether the distribution of a character (state) among species is the result of single versus multiple evolutionary origins. In such a study, including the character of interest can be expected to influence the results in favor of a single origin, particularly when parsimony methods are used to reconstruct the phylogenetic tree. Because parsimony methods minimize total homoplasy, including the character of interest will tend to give results in which the numbers of its own reconstructed origins and losses are reduced relative to reconstructions on trees upon which that character had no influence (Maddison 1990; Maddison and Maddison 1992; Swofford and Maddison 1992). Such reductions will be logically problematic if the results support the hypothesis of a single origin. Because including the character of interest is expected to increase the chances that the results will indicate a single origin, that inference is not particularly cautious, and the practice of including the character of interest constitutes an inappropriate influence (but see below).

Nevertheless, if the results support the alternative hypothesis of multiple origins, then including the character of interest is not logically problematic. Because this practice will tend to influence the results in favor a single origin, a result indicating multiple origins is not called into question by that practice. In this case, the inference of multiple origins is conservative and the influence associated with including the character of interest is appropriate (but see below). For similar reasons, excluding the character of interest will be logically problematic if the results favor multiple origins, but it will not be problematic if they favor a single origin. Thus, with respect to the question of single versus multiple origins, both including and excluding the characters of interest can systematically influence the results. Whether the influence associated with either practice is appropriate or inappropriate, however, depends on which hypothesis is supported.

In other cases, the influence associated with including the characters of interest is not problematic, but for a different reason. Consider, for example, studies investigating the potential correlated evolution of two characters (Felsenstein 1985; Maddison 1990; Pagel 1994). Maddison (1990) pointed out that in his parsimony-based method for analyzing correlated character evolution, including the characters of interest will tend to reduce the number of gains and losses in those characters relative to optimizing them on a tree upon which they had no influence. Therefore, he concluded that including the characters of interest will likely render the test more conservative because with fewer changes it will be more difficult to obtain a significant correlation. For the same reason, excluding the characters of interest will likely render the test more liberal. It should be noted that the systematic influence in this case merely affects the number of observations, rather than specifically favoring a correlation between the characters or the lack thereof. It is also important to note that including the characters of interest—rather than excluding them—is expected to result in a more conservative test.

In still other cases, neither including nor excluding the characters of interest is expected to constitute a problematic influence. Consider an analysis of whether rates of evolution for a particular set of characters have been constant or variable among certain lineages. Although reconstructing the phylogeny either including or excluding the characters of interest may influence the numbers of inferred changes in those characters, neither practice is expected to predispose the analysis to distribute the changes equally over branches of equal temporal duration or to concentrate them in one or a few such branches. Therefore, provided that the methods of tree reconstruction and character optimization do not contain inherent rate biases, the characters of interest can be optimized on a tree derived from an entirely different set of characters, or they can be optimized on a tree based entirely on the characters of interest. Neither practice should compromise the rigor of the analysis by influencing the results in favor of either constant or variable rates.

From the examples presented above, we can conclude that both including and excluding the characters of interest can systematically influence the outcome of an analysis. Whether such an influence is logically problematic, however, depends on the specifics of the situation. In some cases, including the characters of interest constitutes an inappropriate influence; in other cases, excluding those characters constitutes an inappropriate influence; and in still other cases, neither practice constitutes a significant influence. Consequently, inferential rigor cannot be guaranteed by adopting a simple rule to always exclude (or include) the characters of interest. Indeed, adopting a such a rule can sometimes have the opposite effect intended by the researcher—that is, it can result in a less rigorous analysis. Sound inferences require careful consideration of how each practice is expected to influence the results of a given analysis and how those influences bear on the evaluation of the specific hypotheses under consideration.

MORPHOLOGY VERSUS MOLECULES

Concerns about including versus excluding characters are most often raised in reference to morphological characters (and to a lesser extent, behavioral, ecological, and physiological ones). On the one hand, biologists commonly have an interest in the evolution of morphological characters (as well as behavioral, ecological, and physiological ones); on the other hand, morphological characters (unlike behavioral, ecological, and physiological ones) are the data most often available for reconstructing phylogenies—indeed, in many cases they are the only available data. According to biologists who believe that including the characters of interest is logically problematic, the solution is to reconstruct the tree based on an "independent" data set, most often a molecular one (e.g., Hedges and Maxson 1996). But biologists are also interested in the evolution of molecular characters, and one rarely, if ever, hears the argument that phylogenies based on an "independent" morphological data set are needed to analyze the evolution of molecular characters, though this conclusion follows from the premise that including the characters of interest is logically problematic (see also Lee 1997).

This discrepancy may be attributable, at least in part, to a difference in the kinds of questions that are being investigated concerning the evolution of molecular versus morphological characters. For example, studies about the evolution of molecular characters more commonly address issues about rates of evolution, while those about the evolution of morphological characters more commonly address issues about single versus multiple origins. Given this situation and the conclusions (see preceding section) that biases associated with including and excluding the characters of interest generally are not problematic for studies of evolutionary rates, but can be for studies of single versus multiple character state origins, perhaps including the characters of interest presents fewer problems for molecular characters. Nevertheless, the discrepancy also seems to reflect, at least in part, a prejudice on the part of certain researchers against morphological (as well as ecological, physiological, and behavioral) characters relative to molecular ones. For example, though largely unsubstantiated (Sanderson and Donoghue 1989), there is a widespread opinion that morphological characters are more subject to homoplasy than molecular ones (e.g., Hedges and Maxson 1996). But regardless of the relative frequency of homoplasy in different classes of characters, my previous conclusions about the influences associated with including and excluding the characters of interest do not justify different standards for morphological versus molecular characters. If the practice of including (or excluding) the characters of interest introduces an inappropriate influence, then a logical problem will exist regardless of whether the characters of interest are morphological or molecular. Similarly, if the practice of including (or excluding) the characters of interest does not introduce an inappropriate influence, then no logical problem will exist regardless of whether the characters of interest are morphological or molecular.

Another manifestation of the prejudice against morphological characters is the extremely questionable practice of excluding all morphological characters on the grounds that the characters of interest are morphological (e.g., Block and Finnerty 1994; Meyer et al. 1994). As noted above, it may sometimes be justifiable to exclude the characters of interest to avoid an inappropriate influence, and it may even be justifiable to exclude other characters if they are thought to be adaptively (i.e., functionally and phylogenetically) correlated with the characters of interest. For example, someone studying the evolution of pollination systems might want to exclude certain characters, such as stamen length, that are thought to be an integral component of the pollination system. It is not justifiable, however, to exclude characters on the basis of their sharing some property with the characters of interest that has no bearing on the hypothesis or inference in question. Thus, when investigating the evolution of a particular morphological character, it is not justifiable to exclude other characters that are not thought to be adaptively correlated with the character of interest for the irrelevant reason that those characters also happen to be morphological (see also Donoghue and Sanderson 1992).

HOMOLOGS VERSUS ANALOGS

There are some cases in which it is reasonably clear that the characters of interest should not be included as part of the data used to reconstruct the phylogenetic tree. These are cases in which the characters of interest are classes of traits, often defined in terms of function, which are considered analogous rather than homologous prior to the phylogenetic analysis (usually on the grounds of similarity). For example, the class of traits referred to as floral color change may involve the petals in some species, the ovary in others, and the stamens in still others (Weiss 1991). Similarly, the class of traits known as sexual dimorphism includes both cases in which males are larger than females and those in which females are larger than males. To the extent that such classes of traits are considered to consist of non-homologous components, they are likely to be misleading regarding phylogenetic relationships and thus should not be included—at least not as functionally defined classes—in the phylogenetic analysis (McLennan and Brooks 1993). This does not mean, however, that the individual components should not be included. Provided that those components consist of traits that are considered potentially homologous, they might well be included. Thus, although one might not want to include a character with the states "sexual dimorphism absent" and "sexual dimorphism present"; it might nevertheless be acceptable (at least in some cases) to include a character with the states "males larger than females," "males and females of equal size," and "females larger than males." Furthermore, just because the different instances of a functionally defined character are not the same (identical) does not mean that they are not potential homologs. For example, the fact that cooperative breeding in birds includes systems with both male and female helpers, as well as systems with male helpers only, does not mean those systems are necessarily non-homologous (contra McLennan and Brooks 1993). It is at least possible that a system with male and female helpers has been derived from one with male helpers only. All of these considerations apply primarily to studies that seek inferences that are the most likely to be correct. If, on the other hand, the primary concern is inferential caution, it may sometimes be permissible to include a character even when it is thought to include non-homologous components—if doing so makes the basis for an inference more conservative. For example, the inference that floral color change has evolved multiple times within flowering plants would only be strengthened by including the character "floral color change" (i.e., regardless of the specific structures involved) in the phylogenetic analysis, since this practice would tend to bias the analysis against the inference of multiple origins.

DIFFERENT TYPES OF STUDIES

Studies of character evolution can be classified in several ways with regard to some very general properties. Some generate new hypotheses, while others test existing hypotheses. Some evaluate the relative support for two or more alternative hypotheses (e.g., single versus multiple origins), while others attempt to reject a null hypothesis (e.g., no phylogenetic association between characters). Some focus on individual character transformations (e.g., origin of the avian flight apparatus), while others emphasize general classes of character transformations (e.g., origin of cooperative breeding in all taxa in which it occurs) (see McLennan and Brooks [1993] and Edwards and Naeem [1994] for different perspectives on this issue). These different classes of studies have somewhat different consequences for the practices of including and excluding the characters of interest.

On the one hand, studies in many of the categories commonly adopt the perspective described above that seeks the estimate or inference about character evolution that is the most likely to be correct regardless of how much better it is supported than an alternative inference. For example, in proposing an hypothesis about the evolution of a particular character, one might accept the most parsimonious reconstruction of character state transformation inferred from the minimum length tree. Or one might favor whichever hypothesis, single or multiple character state origins, maximized the value of the relevant likelihood function on a given tree. The only studies that seem to be prohibited from adopting this perspective are those that explicitly attempt to reject a null hypothesis—that is, studies involving significance testing. By adopting the convention of rejecting the null hypothesis only when the data have a relatively low probability if the null is true, such studies implicitly adopt the alternative perspective of accepting an inference only if it is unlikely to be incorrect. In any case, under the perspective that seeks the single best estimate or inference about character evolution, it may generally be preferable to include the characters of interest not regardless of, but in that context because of, their potential influence on the results. Provided that the characters of interest contain phylogenetic information, that information is expected to improve the accuracy of inferences about their own evolution.

Even within the context of the perspective that emphasizes inferential caution, there are important differences between generating hypotheses and evaluating the relative merits of alternative hypotheses, on the one hand, and attempting to reject null hypotheses, on the other. When attempting to reject a null hypothesis, that hypothesis is explicitly identified prior to the analysis. Therefore, provided that the influence of excluding or including the characters of interest on the attempt to reject that hypothesis can also be determined prior to the analysis, the practice that renders the test more conservative can simply be chosen (but see below). Of course, failure to reject the null hypothesis should not be interpreted as support for the null hypothesis, since the influence associated with either including or excluding the characters of interest has been exploited to make the test more conservative—that is, to make it more difficult to reject the null hypothesis even if it is false.

In contrast, when a phylogenetic analysis is used to generate, rather than to reject, an hypothesis about character evolution, that hypothesis in many cases will not have been identified prior to the analysis. This will make it impossible to know the influence of either including or excluding the characters of interest on the hypothesis in question until after the analysis has been performed and the hypothesis generated. Similar problems arise in analyses that simultaneously evaluate alternative hypotheses. In such analyses, it may be possible to determine which practice, including or excluding the characters of interest (if either), should tend to favor a particular hypothesis (but see below). Nevertheless, if the hypotheses are strict alternatives—such that contradiction of one implies support for the other and vice versa—then the outcome of the analysis becomes critical. In other words, it may still be impossible to identify the practice that results in a more conservative inference until after the analysis has been conducted, since one must first know which practice favors the inference that is actually supported by the analysis.

RECOMMENDATIONS

Given that both including and excluding the characters of interest can potentially compromise the conservativeness of an analysis, and given that in some cases it is not possible to identify which practice will do so prior to the analysis, how are comparative biologists to achieve logically sound inferences about character evolution? Although this question has not yet received a great deal of attention, several authors have already used methods or approaches that deal with the problem of including and excluding characters in more sophisticated ways. I will next review some these methods in an attempt to provide some practical guidance. My intent is not so much to endorse the particular methods that I discuss but rather to use those methods as examples of a general approach to analyses of character evolution whose conclusions do not suffer from inappropriate influences.

Perhaps the most straightforward approach is to conduct phylogenetic analyses both including and excluding the characters of interest. If the trees resulting from the separate analyses turn out to be identical in terms of whatever properties might affect the inference about character evolution (e.g., topology, branch lengths—though the latter are unlikely to be identical if they are estimated from the data that includes and excludes the characters of interest), then the inclusion versus exclusion of the characters of interest is irrelevant to inferences about their evolution. Evolutionary transformations in the characters of interest are simply reconstructed on the tree and the appropriate inferences about character evolution are drawn from them. On the other hand, if the trees resulting from the separate analyses differ, then transformations in the characters of interest are reconstructed on the alternative trees and inferences about character evolution are drawn from them. If the inferences about character evolution are identical,[2] then the inclusion versus exclusion of the characters of interest is irrelevant, and the results can be accepted. If, on the other hand, the results based on the alternative trees differ, then relationships critical to evaluating the relative merits of the inferences in question are too tenuous for either to be accepted as a conservative inference. A better supported phylogeny and hence more data are needed (see also Swofford and Maddison 1992; Losos 1994).

The approach described above should be useful both for generating hypotheses about character evolution and for evaluating the relative merits of alternative hypotheses. A variant of it was used by Armbruster (1993) in his analysis of pollination and mating system evolution in *Dalechampia,* a clade of neotropical vines. Although Armbruster excluded the characters of interest from his phylogenetic analyses, he was nevertheless concerned about possible correlations between those characters and some of the characters that were used to reconstruct the phylogeny, particularly those describing certain aspects of floral morphology. Therefore, he performed separate phylogenetic analyses including and excluding the morphological characters that he believed might be correlated with pollination and mating systems. He found that the separate analyses yielded similar trees, with the main difference being that the tree for the reduced data set (presumably a consensus tree) had fewer resolved nodes. Several of Armbruster's inferences about the evolution of pollination and mating systems in *Dalechampia* were supported by both analyses, including the inference that pollination systems have been evolutionarily labile. Those inferences apparently are not dependent on including (or excluding) characters of floral morphology that are potentially related to pollination and mating systems.

Another method based on including and excluding characters was used by Ronquist (1994) in his analysis of the evolution of a type of nest parasitism in cynipid wasps. Some species of cynipids, known as inquilines, cannot induce galls in plants; instead, they are nest parasites whose young develop in the galls induced by other species of cynipids, the hosts. Ronquist wished to address the question of whether the inquilines had originated once, presumably from one of their hosts, and later spread to other hosts (inquiline monophyly), or whether each inquiline had originated separately, presumably from its host (inquiline polyphyly). He was concerned, however, that a phylogenetic analysis might be confounded by convergence, either among the inquilines, because of their similar parasitic mode of life, or between parasites and hosts, because of their similar environments. He reasoned that if the characters responsible for grouping

[2]This is not necessarily the same as saying that the reconstructions of individual character transformations are identical—particularly when inferences about character evolution are generalizations about classes of character transformations.

taxa together (whether different parasites or hosts and their parasites) in an initial analysis were convergent, then removal of those characters should result in a fundamental change in relationships (so that, for example, a parasite might first be grouped with its host, but afterwards with other parasites). On the other hand, if the characters were indicative of the true relationships, then their removal should result only in a loss of resolution. Ronquist performed an analysis incorporating this reasoning on 12 taxa of cynipid wasps, 4 of which were inquilines. His initial analysis supported a single origin of the inquilines, and subsequent removal of characters common to all inquilines did not change this result. Support for the hypothesis of inquiline monophyly apparently is not simply the result of including convergent characters related to inquilinism in the analysis.

Because multiple characters are generally used to reconstruct phylogenies, a potential exists for complex interactions among those characters in determining the optimal tree(s). Consequently, it is not always possible to predict the effects of including versus excluding the characters of interest on the outcome of an analysis by simple deduction. A study by Weller et al. (1995) on the evolution of breeding systems in Alsinoideae, a clade of endemic Hawaiian plants, illustrates this problem. Weller et al. used parsimony-based methods to reconstruct phylogenetic relationships among species of Alsinoideae including and excluding a character describing the breeding system, as well as several other characters describing aspects of floral morphology that were thought to be related to the breeding system. They then used the phylogeny and character optimization methods based on parsimony to infer the number of times breeding system dimorphism (gynodioecy, subdioecy, dioecy) had evolved. One might expect that exclusion of the breeding system character from the phylogenetic analysis and optimizing it on the tree afterwards would lead to inferring an equal or greater number of origins for dimorphism than optimizing it on a tree based on a data set that included the character. However, Weller et al. found that some of the most parsimonious trees required more origins of dimorphism when the breeding system character was included than when it was excluded. This result was obtained in at least some cases under both accelerated and delayed transformation methods of character optimization and treating the character as either ordered or unordered. A similar result was obtained by Armbruster (1993) regarding the consistency index of his pollination system character, which was higher (indicating less rather than more homoplasy) when the characters thought to be related to it were excluded from the phylogenetic analysis. These examples indicate that the effects of including versus excluding characters may sometimes run counter to expectations, suggesting that it is advisable to examine those effects empirically.

The analysis of character evolution on alternative trees can be extended from the optimal trees based on the analysis of data sets including and excluding the characters of interest to larger sets of trees—such as all the resolved trees derivable from a strict consensus tree of the optimal trees for data sets including and excluding the characters of interest, or a random sample of those trees. To the extent that additional trees allow for additional reconstructions of character evolution, this approach can be used to increase the conservativeness of an analysis, though it might also be expected to decrease analytical power (i.e., the ability to make an inference). Losos (1994) and Martins (1996) have proposed using computer simulations to generate sets of trees for unresolved parts of a phylogeny, and Martins (1996) has described methods for estimating the parameter of interest, as well as confidence intervals, on sets of alternative trees.

When attempting to reject a null hypothesis, it is sometimes possible to estimate both the phylogenetic tree and the evolution of certain characters simultaneously, using the null hypothesis as a constraint in the estimation process. This approach was suggested with regard to testing for correlated character evolution by Felsenstein (1985) and Pagel (1994) and has recently found broad application in comparative phylogenetic studies (Huelsenbeck and Rannala 1997). In the approach described by Felsenstein and by Pagel, the characters of interest are included in the data used to estimate the tree, and the optimality criterion, in this case the likelihood, is a single multivariate quantity describing both the tree and the parameter of interest (in this case, the correlations). The maximum value of the likelihood function with the characters of interest unconstrained (allowing for correlated evolution) is then compared, in terms of a statistical criterion for significance, with the maximum likelihood obtained when the characters of interest are constrained to correspond with the null hypothesis (independent evolution). The significance testing approach guarantees conservativeness in terms of the probability of rejecting a true null hypothesis, but the effect of simultaneously optimizing the character correlations and the tree, as opposed to optimizing the character correlations on a tree estimated from other characters is not obvious. Once again, the most straightforward way to determine that effect is to do the analysis both ways.

CONCLUSIONS

Sometimes a distinction is made between systematics, the discipline concerned primarily with reconstructing or estimating phylogenetic trees, and comparative biology, the discipline concerned primarily with using those reconstructed trees to study character evolution. Deleporte (1993), for example, categorized the biologists associated with those respective disciplines as "phylogeny makers" and "phylogeny users." Although this dichotomy is useful for describing different research emphases, it is also oversimplified. One thing that should be clear from the preceding analysis is that the rigorous evaluation of hypotheses about character evolution often demands the reconstruction or estimation of phylogenetic trees for that express purpose. In other words, comparative biologists cannot afford to limit themselves to using trees generated by systematists; they need to be actively involved in tree reconstruction. This need is obvious in cases where no phylogeny or only a poorly resolved one is available; but the present analysis indicates that it is also important for achieving cautious or conservative inferences even when a resolved phylogeny is available. The benefits to comparative biologists of being versed in the analysis of systematic data—or alternatively, of working closely with systematists—are tremendous. Comparative biologists who are able to produce their own trees will be able to satisfy their own unique needs, from investigating the effects of including and excluding characters (e.g., Armbruster 1993; Ronquist 1994; Weller et al. 1995), to generating sets of trees for unresolved parts of phylogenies (e.g., Losos 1994; Martins 1996), to comparing optimal unconstrained trees with those that are constrained to conform with various null hypotheses (e.g., Felsenstein 1985; Pagel 1994). But the benefits of generating trees for the express purpose of evaluating evolutionary hypotheses are not limited to promoting more rigorous analyses in this area. One of the greatest benefits will be a more thorough integration of systematics and comparative biology—two closely related fields that have much to gain from one another.

ACKNOWLEDGMENTS

This chapter has its origins in a dissertation that was submitted in partial fulfillment of the requirements of a doctoral degree in zoology at the University of California at Berkeley (K. de Queiroz 1989). I thank my adviser, David B. Wake, and the other members of my committee, Marvalee H. Wake and Vincent M. Sarich, for their comments on the dissertation. Carole C. Baldwin, Alan de Queiroz, Michael J. Donoghue, Joseph Felsenstein, Theodore Garland Jr., Harry W. Greene, Jonathan B. Losos, Emilia P. Martins, Molly R. Morris, and John W. Wenzel provided further discussions or comments. I would also like to thank John J. Wiens for inviting me to participate in the symposium of the Society of Systematic Biologists and to contribute a chapter to this book, as well as for his comments on the manuscript.

REFERENCES

Armbruster, W. S. 1992. Phylogeny and the evolution of plant–animal interactions. Bioscience 42: 12–20.

Armbruster, W. S. 1993. Evolution of plant pollination systems: Hypotheses and tests with the neotropical vine *Dalechampia.* Evolution 47: 1480–1505.

Block, B. A., and J. R. Finnerty. 1994. Endothermy in fishes: A phylogenetic analysis of constraints, predispositions, and selection pressures. Environmental Biology of Fishes 40: 283–302.

Brooks, D. R., and D. A. McLennan. 1991. Phylogeny, Ecology, and Behavior: A Research Program in Comparative Biology. University of Chicago Press, Chicago.

Carpenter, J. M. 1989. Testing scenarios: Wasp social behavior. Cladistics 5: 131–144.

Coddington, J. A. 1988. Cladistic tests of adaptational hypotheses. Cladistics 4: 3–22.

Deleporte, P. 1993. Characters, attributes, and tests of evolutionary scenarios. Cladistics 9: 427–432.

de Queiroz, A., and P. H. Wimberger. 1993. The usefulness of behavior for phylogeny estimation: Levels of homoplasy in behavioral and morphological characters. Evolution 47: 46–60.

de Queiroz, K. 1989. Morphological and biochemical evolution in the sand lizards. Doctoral dissertation, University of California, Berkeley.

de Queiroz, K. 1996. Including the characters of interest during tree reconstruction and the problems of circularity and bias in studies of character evolution. American Naturalist 148: 700–708.

Donoghue, M. J. 1989. Phylogenies and the analysis of evolutionary sequences, with examples from seed plants. Evolution 43: 1137–1156.

Donoghue, M. J., and M. J. Sanderson. 1992. The suitability of molecular and morphological evidence in reconstructing plant phylogeny. Pages 340–368 *in* Molecular Systematics of Plants (ed. P. S. Soltis, D. E. Soltis, and J. J. Doyle). Chapman and Hall, New York.

Edwards, S. V., and S. Naeem. 1994. Homology and comparative methods in the study of avian cooperative breeding. American Naturalist 143: 723–733.

Eggleton, P., and R. I. Vane-Wright, eds. 1994. Phylogenetics and Ecology. Linnean Society Symposium Series, No. 17. Academic Press, London.

Eldredge, N., and J. Cracraft. 1980. Phylogenetic Patterns and the Evolutionary Process. Columbia University Press, New York.

Farris, J. S. 1983. The logical basis of phylogenetic analysis. Pages 7–36 *in* Advances in Cladistics, Volume 2: Proceedings of the Second Meeting of the Willi Hennig Society (ed. N. I. Platnick and V. A. Funk). Columbia University Press, New York.

Felsenstein, J. 1978. Cases in which parsimony or compatibility methods will be positively misleading. Systematic Zoology 27: 401–410.

Felsenstein, J. 1985. Phylogenies and the comparative method. American Naturalist 125: 1–15.

Funk, V. A., and D. R. Brooks. 1990. Phylogenetic systematics as the basis of comparative biology. Smithsonian Contributions to Botany 73: 1–45.

Harvey, P. H., and M. D. Pagel. 1991. The Comparative Method in Evolutionary Biology. Oxford University Press, Oxford.

Harvey, P. H., and A. Purvis. 1991. Comparative methods for explaining adaptations. Nature 351: 619–623.

Harvey, P. H., A. J. Leigh Brown, and J. Maynard Smith, eds. 1995. New uses for new phylogenies. Philosophical Transactions of the Royal Society of London B 349: 1–118.

Hedges, S. B., and L. R. Maxson. 1996. Re: Molecules and morphology in amniote phylogeny. Molecular Phylogenetics and Evolution 6: 312–314.

Hendy, M. D., and D. Penny. 1989. A framework for the quantitative study of evolutionary trees. Systematic Zoology 38: 297–309.

Hillis, D. M. 1987. Molecular versus morphological approaches to systematics. Annual Review of Ecology and Systematics 18: 23–42.

Huelsenbeck, J. P., and B. Rannala. 1997. Phylogenetic methods come of age: Testing hypotheses in an evolutionary context. Science 276: 227–232.

Kluge, A. G. 1989. A concern for evidence and a phylogenetic hypothesis of relationships among *Epicrates* (Boidae, Serpentes). Systematic Zoology 38: 7–25.

Lauder, G. V., and K. F. Liem. 1989. The role of historical factors in the evolution of complex organismal functions. Pages 63–78 *in* Complex Organismal Functions: Integration and Evolution in Vertebrates (ed. D. B. Wake and G. Roth). John Wiley and Sons, New York.

Lee, M. S. Y. 1997. Re: Molecules, morphology, and phylogeny: A response to Hedges and Maxson. Molecular Phylogenetics and Evolution 7: 394–395.

Losos, J. B. 1994. An approach to the analysis of comparative data when a phylogeny is unavailable or incomplete. Systematic Biology 43: 117–123.

Maddison, W. P. 1990. A method for testing the correlated evolution of two binary characters: Are gains and losses concentrated on certain branches of a phylogenetic tree? Evolution 44: 539–557.

Maddison, W. P., and D. R. Maddison. 1992. MacClade: Analysis of Phylogeny and Character Evolution. Sinauer, Sunderland, Mass.

Martins, E. P. 1996. Conducting phylogenetic comparative studies when the phylogeny is not known. Evolution 50: 12–22.

McKey, D. 1991. Phylogenetic analysis of the evolution of a mutualism: *Leonardoxa* (Caesalpiniaceae) and its associated ants. Pages 310–334 *in* Ant–Plant Interactions (ed. C. R. Huxley and D. F. Cutler). Oxford University Press, Oxford.

McKitrick, M. C. 1993. Trends in the evolution of hindlimb musculature in aerial-foraging birds. Auk 110: 189–206.

McLennan, D. A., and D. R. Brooks. 1993. The phylogenetic component of cooperative breeding in perching birds: A commentary. American Naturalist 141: 790–795.

Meyer, A., J. M. Morrissey, and M. Schartl. 1994. Recurrent origin of a sexually selected trait in *Xiphophorus* fishes inferred from a molecular phylogeny. Nature 368: 539–542.

O'Hara, R. J. 1988. Homage to Clio, or, toward an historical philosophy for evolutionary biology. Systematic Zoology 37: 142–155.

Olmstead, R. 1989. Phylogeny, phenotypic evolution, and biogeography of the *Scutellaria angustifolis* complex (Lamiaceae): Inference from morphological and molecular data. Systematic Botany 14: 320–338.

Pagel, M. 1994. Detecting correlated evolution on phylogenies: A general method for the comparative analysis of discrete characters. Proceedings of the Royal Society of London B 255: 37–45.

Patterson, C., ed. 1987. Molecules and Morphology in Evolution: Conflict or Compromise? Cambridge University Press, Cambridge.

Ronquist, F. 1994. Evolution of parasitism among closely related species: Phylogenetic relationships and the origin of inquilinism in gall wasps (Hymenoptera, Cynipidae). Evolution 48: 241–266.

Sanderson, M. J., and M. J. Donoghue. 1989. Patterns of variation in levels of homoplasy. Evolution 43: 1781–1795.

Sillén-Tullberg, B., and A. P. Møller. 1993. The relationship between concealed ovulation and mating systems in anthropoid primates: A phylogenetic analysis. American Naturalist 141: 1–25.

Swofford, D. L., and W. P. Maddison. 1987. Reconstructing ancestral character states under Wagner parsimony. Mathematical Biosciences 87: 199–229.

Swofford, D. L., and W. P. Maddison. 1992. Parsimony, character-state reconstructions, and evolutionary inferences. Pages 186–223 *in* Systematics, Historical Ecology, and North American Freshwater Fishes (ed. R. L. Mayden). Stanford University Press, Stanford, Calif.

Sytsma, K. J. 1990. DNA and morphology: Inference of plant phylogeny. Trends in Ecology and Evolution 5: 104–110.

Vane-Wright, R. I., S. Schulz, and M. Boppré. 1992. The cladistics of *Amauris* butterflies: Congruence, consensus, and total evidence. Cladistics 8: 125–138.

Weiss, M. R. 1991. Floral colour changes as cues for pollinators. Nature 354: 227–229.

Weller, S. G., W. L. Wagner, and A. K. Sakai. 1995. A phylogenetic analysis of *Schiedea* and *Alsinidendron* (Caryophyllaceae: Alsinoideae): Implications for the evolution of breeding systems. Systematic Botany 20: 315–337.

INDEX

Page-number citations followed by *f, t,* or *n* refer to figures, tables, or notes appearing on those pages.

N

O

P

Q

R

S

T

U

V

W